SOUTHERN OVERBERG

Protea repens

Following page: Ornithogalum dubium *at Zoetendals Vallei.*

SOUTHERN OVERBERG

South African Wild Flower Guide 8

Text by:
Penny Mustart, Richard Cowling and Janice Albertyn

Photography by: Colin Paterson-Jones

This guide is the 8th in the Botanical Society's series of Wild Flower Guides and is published jointly by the Botanical Society and the Institute for Plant Conservation in association with the National Botanical Institute

The Botanical Society of South Africa which was founded in 1913 supports the National Botanic Gardens, promotes the conservation and cultivation of our indigenous flora and provides environmental education.

One of our projects is the publication of a series of wild flower guides.

Published to date are:

Guide 1: Namaqualand and Clanwilliam	1981 (out of print)
Guide 1: (revised): Namaqualand	1988/1996/1997
Guide 2: Outeniqua, Tsitsikamma & Eastern Little Karoo	1982/1997
Guide 3: Cape Peninsula	1983/1996
Guide 4: Transvaal Lowveld & Escarpment	1984
Guide 5: Hottentots Holland to Hermanus	1985
Guide 6: Karoo	1994/1997
Guide 7: West Coast	1996

Opposite title page: Drosera cistiflora

First edition, first impression 1997
Botanical Society of South Africa
Kirstenbosch, Claremont 7735 RSA

© Copyright: Text: University of Cape Town
© Copyright: Photographs: Colin Paterson-Jones

Design, typesetting and production by Wim Reinders
Reproduction by CMYK pre-press, Cape Town
Printed and bound by National Book Printers, Drukkery Street, Goodwood, Cape

All rights reserved. No part of this publication may be reproduced, stored in a retrieval system or transmitted, in any form or by any means, electronic, mechanical, photocopying, recording or otherwise, without the permission of the copyright owner.

ISBN 1-874999-15-5

Foreword

Welcome to a long-awaited popular guide to the Cape Floral Kingdom of the Southern Overberg - an undiscovered but diverse region literally over the mountains (Over't Geberghte) to the south east of Cape Town. Here, at the tip of the African continent, the long rib-cages of limestone thrust up their spines, washed by brisk breezes from the southern oceans and clad in spectacular endemic plants such as the Bredasdorp sugarbush (*Protea obtusifolia*) and a wealth of cone bushes or tolbosse (*Leucadendron* species). The mountain slopes are graced by jewels such as the tiny *Protea denticulata* and the famed Bredasdorp lily (*Cyrtanthus guthrieae*). A diversity of coastal habitats further enhances this amazing display of nature's generosity.

This book will be eagerly received by professional conservationists and equally enthusiastically by the many amateur botanists and other Overbergers who have this unique vegetation at the heart of their being. Who better to produce this guide than long-standing fynbos champions Penny Mustart, Richard Cowling and Janice Albertyn and their many committed supporters? Superb photography by Colin Paterson-Jones sets the seal on this celebration of the Cape Floral Kingdom, the eighth fynbos guide in the Botanical Society's unsurpassed series.

There is place for people in the fynbos of the Overberg. Here you will encounter the legendary hospitality among landowners who cherish this heritage, many of whom are exploring the possibilities of fynbos-based ecotourism. Carefully planned facilities for a specialized fynbos experience for small groups of discerning travellers holds the promise of providing benefits to tourists, local communities and conservation alike.

We invite you to explore this little-known gem and linger among the honey-scented blombos bushes. Meet the birds and insects and other animals, large and small, that depend on this vegetation. And above all, enjoy and cherish what remains of this unique and spectacular form of plant life.

Ann Scott
Overberg Conservation Services
De Kelders
March, 1997

Contents

Foreword   7
Acknowledgements   11
Why this book   15
History and geography of the Southern Overberg   15
    History   15
    Scenery and soils   16
    Climate and weather   17
Flora and vegetation   20
    Flora   20
    Vegetation   20
        Acid sand proteoid fynbos   21
        Limestone proteoid fynbos   22
        Neutral sand proteoid fymbos   24
        Ericaceous fynbos   25
        Dune asteraceous fynbos   26
        Elim asteraceous fynbos   26
        Wet restioid fynbos   26
        Dry restioid fynbos   28
        Renosterveld   31
        Forest and thicket   31
        Coastal strand and rocky shelf vegetation   31
        Wetlands   32
        Alien vegetation   33
Conservation and land use   35
Choice and arrangements of plates   35
Companion books   36

DESCRIPTION OF PLANTS   38

Glossary   254
Index   257

Spiloxene capensis

INSTITUTE FOR PLANT CONSERVATION

These guides are produced through the co-operation of members of the public and private sector and their dedication to the conservation of our floral wealth. The publication of this volume has been made possible by contribution from the following bodies: the Botanical Society of South Africa, the National Botanical Institute, Kirstenbosch and the Institute for Plant Conservation, University of Cape Town. The Publications Committee that has motivated the series consists of representatives from the Botanical Society of South Africa, Cape Nature Conservation and the National Botanical Institute.

Acknowledgements

This book is a tribute to the people and plants of the Southern Overberg. It acknowledges people, past and present, who have been instrumental in recording this special flora - amateurs who collected plants in the field, scientists who named them and herbarium staff who curated the specimen collections. This book also acknowledges the role played by the custodians of the region's flora and vegetation: staff of De Hoop and other nature reserves; landowners proud of their priceless heritage and especially those who have taken the step to proclaim and manage their land as private nature reserves; and the people of Elim who are restoring their unique fynbos on Geelkop.

The authors and photographer wish to acknowledge the help that they received from the following people:

Joy Abrahams, Christine Afrika, Johan and Margaret Albertyn, Pieter and Ann Albertyn, Rory Allardice, Carolina Appollis, Anne Bean, Ian and Avanol Bell, Erna Bredenkamp, Heynè and Sharon Brink, Chris Burgers, Pietman Cillie, Hannie Cloete, Mick and June D'Alton, Thys de Villiers, Diana Durrant, Caroline Engel, Trevor Farr, Peter Goldblatt, Jenny Helmsley, Barry Heydenrych, Giel Huga, Roelof Jalving, Jessica Kemper, Gerhard Kirsten, Peter Linder, Mandy Lombard, John Manning, Christo Marais, Bertus Meyer, John Mustart, Ted and Inge Oliver, Shirley Pierce, John Rourke, Mike and Ann Scott, Jurgens Smit, Dee Snijman, Paul Swart, Petro Swart, Test Flight and Development Centre, Etienne van Blerk, Muriel van Breda, Eddie and Diana van Heerden and Ernst van Jaarsveld.

In particular, Janice wishes to thank her wonderful mother-in-law, Jewel Albertyn, for all she taught her about the plants of this area and for what she did to conserve them. Her personally collected and curated herbarium has been of use to many plant-lovers.

The role of the Botanical Society and the Pew Charitable Trusts in underwriting the production of this book is acknowledged. The support and material help of John Albertyn is gratefully acknowledged. We also thank Wim Reinders for designing the guide and overseeing its production and Dick Geary-Cooke who prepared the index.

Why this book?

Many guides to the Cape Flora have been published over the past two decades – why another and why, specifically, for the Southern Overberg? Firstly, the Southern Overberg, an area of some 3 600 km² along the south-western Cape coast, is the heart of the lowland fynbos, an ecosystem of immense floral biodiversity but increasingly disappearing under the onslaught of alien plants, agriculture and coastal resort development. Secondly, the region is home to the only substantial reserve network in lowland fynbos: De Hoop Nature Reserve, De Mond Nature Reserve and several private nature reserves. Other conservation areas are Salmonsdam Nature Reserve, located in mountain fynbos and the pending Agulhas National Park. With the growth of ecotourism in the Overberg, increasing numbers of people are visiting these reserves in order to appreciate their rich floras. Thirdly, many landowners harvest wildflowers commercially and others are turning to ecotourism to supplement their incomes; for them, a guide to the region's wildflowers is a practical asset.

History and geography of the Southern Overberg

History

Humans have lived in the Overberg for at least half a million years. The early hunter-gatherer inhabitants were dependent on the region's indigenous plants and animals to sustain their lifestyles.

De Kelders, near Gansbaai, has provided some of the earliest evidence of livestock in the Western Cape: sheep bones dating to about 2 000 years ago have been found there. The Khoi-khoi, nomadic pastoralists, were constantly seeking new grazing sites for their sheep and cattle as the old ones became depleted. Thus the short and nutritious grasslands of the Zoetendals Vallei floodplain were grazed by livestock in the company of bontebok and hippopotami.

After the arrival of Europeans at Table Bay in 1652 the activities of the Dutch East India Company came into conflict with the lifestyle of the Khoi-khoi and eventually led to their dispersal. As the colonial frontier expanded loan farms were established throughout the Overberg and in 1747 a magistracy (Drostdy) was established at the site of present-day Swellendam. In time the improvement

Previous page: *Anemone tenuifolia*
Opposite: *Chondropetalum microcarpum*

Early settler cottage and 'werf' at Renosterkop, near Cape Agulhas

of transport networks led to the birth of new settlements: Bredasdorp was founded by Michiel van Breda in 1838. Earlier, van Breda had successfully pioneered the establishment of merino sheep on his farm Zoetendals Vallei, near Cape Agulhas. Other landmarks in the 19th Century history of the Southern Overberg were the establishment by the Moravians of the Elim Mission in 1824 and the completion of the Cape Agulhas lighthouse in 1848.

The 20th Century has seen many changes sweeping across the plains of the Southern Overberg. After the 1940s the widespread introduction of mechanised agriculture saw a shift from veld-based grazing to cereal crops and artificial pastures. Much of the natural renosterveld on the more fertile clayey soils has thus disappeared. The development of holiday resorts and the widespread expansion of thickets of Australian wattles since the 1960s has changed the face of much of the coastline.

Scenery and soils

The landscapes of the Southern Overberg are part of the Cape Folded Belt, a band of parallel ranges of quartzitic sandstone separated by undulating valleys of shale. The first deposits were laid down about 450 million years ago and the mountains were formed some 200 million years later. Other than changes in sea level and some tilting of the coastal plain, Overberg scenery has remained unchanged for the past 65 million years.

The soils of this ancient landscape are mostly acidic, sandy and highly infertile. This is especially true of the sandstone-derived soils of the hills and wind-blown sands in the west. However, most of the Southern Overberg is a low coastal plain covered by sands of marine origin. The oldest coastal deposits are the Bredasdorp limestone formation, rocks deposited between 25 and 10 million years ago when sea levels were much higher than at present. Today these are clearly visible as the limestone ridges south (e.g. Heuningrug) and east of Bredasdorp. Younger deposits include a wide band of alkaline, wind-blown sands (rietveld), the muds and sands of the Zoetendals Vallei and the white, coarse-grained sands and dune rock formations (prominent at Koppie Alleen in the De Hoop Nature Reserve) of the contemporary coastline. An unusual feature of the low, shale-derived valleys in the west is the preservation of large areas of ferricrete (or koffieklip), remnants of the weathering of clayey soils during the same period when the limestones were deposited. Today these surfaces support Elim fynbos, a vegetation type entirely confined to the Southern Overberg.

Climate and weather

The Southern Overberg is an area transitional between the winter-rainfall region in the west and the non-seasonal rainfall region in the east. During the winter months cold fronts and their associated westerly winds bring moderate amounts of drenching rain, especially in the west between Stanford and Cape Agulhas. During summer, but especially in spring and autumn, the ridging South Atlantic high, cut-off lows and southerly air streams result in summer showers. These warmer season rains are most frequent east of Cape Agulhas, where the sea breezes are drawn across the warm Agulhas Current. Thus, the western areas get up to 70% of their rain during winter, whereas in the east, almost half the rain falls between late spring and early autumn.

Overall, the Southern Overberg is a relatively dry area. Annual rainfall along the coast ranges from about 600 mm near Stanford to less than 400 mm at the mouth of the Breede River, east of Potberg. The mountains are much wetter: the western part of the Bredasdorpberge may receive more than 800 mm annually. Temperatures are moderate. The summer days are warm (20-30 $^{\circ}$C), but this is also the time of relentless south-easterly gales. Winter days are cool (12-18 $^{\circ}$C) yet often sun-filled in the periods between the fronts. Winter nights may be cold but freezing conditions are very rare. Indeed, some of the warmest weather is associated with hot, northerly Berg winds during late winter. Generally, the spring months are somewhat windy. October and November may be cloudy with strong winds, switching rapidly from south-west to south-east. The best time, weather-wise, is autumn – between late March and late June the days are often clear, still and warm.

Following page: Haemanthus coccineus *west of Bredasdorp.*

Flora and vegetation

Flora
Like most areas of the Cape Floristic Region, the Southern Overberg has a rich flora – estimated at about 2 500 species. About 300 of these are restricted to the area (*i.e.* endemic), many occurring in very small and scattered populations. It is no surprise, therefore, that the Southern Overberg is home to some 32 species which are variously threatened with extinction. Most of the endemics are associated with limestone soils, and the gravelly clays centred on Elim. Many of the Red Data Book species owe their status to threats from alien plants and agriculture. Others are naturally rare.

Rare and endemic species are often clustered in so-called hot-spots. One such hot-spot is the Groot Hagelkraal area near Pearly Beach. Here an area of limestone, some 5 km^2 in extent, is the exclusive home to six species. The farm supports another 21 species, more or less restricted to a triangle defined by Danger Point, Cape Agulhas and Elim. Such a concentration of endemic plants is without parallel, not only elsewhere in the Cape Floristic Region, but in the world. Other hot-spots are the Elim area, De Poort (now sadly degraded), Soetanysberg hills and adjacent flats and the Potberg range and seaward flats.

The ten largest families in the flora are, in decreasing order of size, Asteraceae, Fabaceae, Iridaceae, Ericaceae, Mesembryanthemaceae, Campanulaceae, Cyperaceae, Restionaceae, Proteaceae and Rutaceae. The ten largest genera are *Erica*, *Aspalathus*, *Crassula*, *Senecio*, *Muraltia*, *Phylica*, *Gladiolus*, *Thesium*, *Pelargonium* and *Cliffortia*. With the exception of the relatively high rank of the Mesembryanthemaceae ("vygies"), the overall composition of the Southern Overberg flora is typical of coastal lowland and montane sites in the Cape Floristic Region.

Vegetation
The vegetation of the Southern Overberg has been relatively well studied. The distribution of the various vegetation types is shown in the inside cover map. We also present a list of characteristic, but not necessarily the most conspicuous, species for each vegetation type. All of the species listed are illustrated in this book.

For convenience we divide the species into groups or plant types, each defined in terms of their different ecological functions. Bulbs or geophytes are plants which usually die back to an underground storage organ. Restioids are wiry, evergreen, grass-like plants belonging to the Restionaceae. The presence of restioids is the feature that effectively distinguishes fynbos from other vegetation types in the Cape Floristic Region. Restioids must not be confused with sedges (Cyperaceae), another group of evergreen, grass-like plants. Fire ephemerals are short-lived plants that are most prolific in the first few years after a veld fire. They are often inconspicuous in older (more than 10 years old) veld, and persist, until the next fire, as soil-stored seeds. Ericoids are shrubs with hard, small and often tightly rolled leaves. They comprise most of the species in the Southern Overberg flora, including ericas, phylicas and many Proteaceae. Proteoids are large, bushy members of the Proteaceae (mainly species of *Leucadendron* and *Protea*) that form the tallest layer in many fynbos vegetation types. Most proteoids are conspicuous and true to habitat; therefore, they are useful plants for identifying different vegetation types. Trees and thicket shrubs are largely associated with patches of scrub-forest but also occur as scattered individuals in many fynbos vegetation types. Unlike fynbos plants, thicket shrubs are dependent on long fire-free intervals for seedling establishment and many species bear intermittent fruit crops that are dispersed by birds.

Not all the plants illustrated in this book fit neatly into the plant types described above. However, the relative abundance of species belonging to one or more of these types is important for defining the various vegetation types of the Southern Overberg and elsewhere in the Cape Floristic Region. For example fynbos is distinguished by the presence of restioids, often in conjunction with proteoids and ericoids belonging to the Ericaceae. Renosterveld lacks restioids and proteoids and is dominated by ericoid shrubs. Forest and thicket is comprised exclusively of trees and thicket shrubs. The abundance of the different plant types is used to define the vegetation types described below.

Acid sand proteoid fynbos

Acid sand proteoid fynbos occurs widely in the Southern Overberg. Like all types of proteoid fynbos, it has a layer of relatively tall proteoid shrubs. Throughout its range acid sand proteoid fynbos grows on infertile (and acidic) sands derived from Table Mountain Group (TMG) rocks. Several sub-types can be recognised, each distinguished by different dominant proteoid shrubs and their habitat preferences.

Protea compacta is the characteristic proteoid on the deep, acid sands that prevail in the south-western part of the Southern Overberg. Many other plants are confined to this habitat, including the ericoids *Agathosma serpyllacea*, *Erica filipendula*, *E. nudiflora*, *E. regia*, *Lachnaea aurea*, *Leucospermum prostratum* and *Spatalla squamata*. Restioids may be very common, especially in poorly drained sites and in the first few years after fire. Some characteristic species are *Ceratocaryum argenteum*, *Hypodiscus argenteus*, *Mastersiella digitata* and *Staberoha multispicula*. Geophytes are not a conspicuous feature of this vegetation. However, the rare *Gladiolus meridionalis* and widespread *Tritoniopsis dodii* can be seen in the autumn months.

The rocky, shallow soils of the Bredasdorpberge and Potberg hills support a different form of acid sand proteiod fynbos. Here the characteristic proteoids are *Protea longifolia*, *Leucospermum cordifolium* and *Leucadendron platyspermum* in the west and *Protea neriifolia* and *Protea aurea* ssp. *potbergensis* on the Potberg. Other Proteaceae definitive of this vegetation are *Leucospermum truncatulum* and the dwarf *Protea denticulata* (a Potberg endemic). Some typical ericoids are *Adenandra gummifera* (Potberg endemic), *Erica ampullacea*, *E. grisbrookii* and *Muraltia collina*. *Lobostemon sanguineus* is a striking component of the Bredasdorpberge flora. Fire ephemerals are conspicuous in young fynbos, especially on the more fertile lower slopes. These include many species of *Aspalathus*, *Anaxeton virgatum*, and the beautiful Potberg endemic, *Roella rhodantha*. Geophytes are abundant, especially after fire. An exquisite species endemic to the eastern Bredasdorpberge is the Bredasdorp lily, *Cyrtanthus guthrieae*.

Owing to higher summer rainfall, the colluvial and relatively fertile lower slopes of the Potberg support a vegetation similar to the grassy fynbos of the Eastern Cape. The most abundant grass is rooigras (*Themeda triandra*), a tropical species. Other species include the widespread restioid, *Restio triticeus*, and the eastern proteoid, *Leucospermum cuneiforme*.

Limestone proteoid fynbos

The Southern Overberg is the centre of limestone fynbos, an endemic-rich vegetation associated with the Bredasdorp Formation limestones. Soils are alkaline, organic-rich and usually confined to small pot holes in the limestone pavement. Colonisation of this substratum by acid-loving fynbos plants was a major physiological challenge, resulting in the evolution of many limestone-specific species.

Throughout its distribution, from Stanford in the west to the Gouritz River in the east, limestone fynbos is characterised by the presence of *Protea obtusifolia* and *Leucadendron meridianum*. In the wetter, western areas (Hagelkraal and Soetanysberg hills), other typical proteoids include *Leucospermum patersonii* and *Mimetes saxatilis*. *Leucospermum truncatum* and *Leucadendron muirii* are representative of the drier east. The restioid component of limestone fynbos is not particularly diverse. *Thamnochortus paniculatus* predominates in the east and the graceful *Ischyrolepis leptoclados* is widespread. *Ficinia truncata* is an interesting sedge, endemic to limestone fynbos.

The ericoid component includes many limestone endemics, especially among the Rutaceae. Species such as *Diosma guthriei*, *D. haelkraalensis*, *Euchaetes longibracteata* and *E. meridionalis* each have their own unique scent but collectively contribute to the fragrance of the limestone landscape. Surprisingly, for they are generally strongly acid-loving plants, ericas are very common in limestone fynbos. Some conspicuous species are *Erica calcareophila*, *E. mariae*, *E. propinqua* and *E. spectabilis*. Other typical ericoids include *Jamesbrittenia calciphila*, *Metalasia calcicola*, *Muraltia lewisae*, *Phylica selaginoides* and *Euryops linifolius*.

Acid sand proteoid fynbos on the Heuningberg, west of Bredasdorp. Dominant shrubs are Protea compacta *and* Leucadendron xanthoconus.

Limestone proteoid fynbos at De Hoop. Leucadendron meridianum *is on the right and* Protea obtusifolia *on the left.*

Harvesting of the locally endemic Erica irregularis *in neutral sand proteoid fynbos near Gansbaai. The large shrub in the background is* Leucadendron coniferum.

Species that are most apparent in the early post-fire years include the fire ephemerals *Aspalathus calcarea* (and many other species), *Indigofera brachystachya*, *Osteospermum subulatum* and *Syncarpha argyropsis* and low sprouting shrubs such as *Hermannia concinnifolia* and *H. trifoliata*. Characteristic limestone geophytes are *Freesia elimensis*, *Lachenalia muirii* and *Watsonia fergusoniae*.

Neutral sand proteoid fynbos

Occurring as an apron at the base of limestone outcrops, neutral sand proteoid fynbos is associated with deep, limestone-derived soils from which much of calcium has been leached. *Protea susannae* is the proteoid that typifies this vegetation type. Other proteoids are *Leucadendron coniferum* in the west and *Leucospermum fulgens*, restricted to the sand-filled valley between the Potberg and the limestone hills. The prostrate *Leucospermum pedunculatum* is a very common member of this vegetation in the coastal strip between the Soetanysberg and Gansbaai.

Characteristic ericoids include *Erica discolor, E. irregularis, E. lineata, E. rhopolantha* and *Spatalla ericoides* (Hagelkraal endemic). Showy geophytes more or less restricted to neutral sand proteoid fynbos include *Lachenalia bulbifera* and

Erica tenella *is the conspicuous shrub in this patch of ericaceous fynbos on the hills above Napier.*

Bobartia longicyma. Many of the other species in this vegetation type also grow in limestone and acid sand proteoid fynbos.

Ericaceous fynbos

In the Southern Overberg, ericaceous fynbos is restricted to the upper, south-facing slopes of the Bredasdorpberge and the steep, coast-facing slopes of the coastal hills (Buffeljagsberge, Soetanysberg, Potberg). These sites are the wettest in the area, receiving more rain than the adjacent lower-lying areas, plentiful precipitation from summer mist and coastal fog, and less radiation than slopes of different aspect. Like all forms of this vegetation, the rocky, sandstone-derived soils are relatively rich in organic matter but are exceedingly leached and nutrient-poor.

Ericaceous fynbos is characterised by, amongst others, a high cover of *Erica* species, the presence of broad-leaved sedges, and the presence of species belonging to the endemic fynbos families, Penaeaceae, Grubbiaceae and Retziaceae. In the Southern Overberg this vegetation lacks the diversity of ericas found in the magnificent upland heathlands of the Riversonderend and Langeberg mountains to the north.

Many of the ericas found in acid sand proteoid fynbos also grow in ericaceous

Dune asteraceous fynbos at De Hoop. A lone Thamnochortus insignis *plant is on the right and* Sideroxylon inerme *occupies the left background. The pink-flowered shrub is* Acmadenia macropetala.

fynbos. Some distinctive species include *Erica colorans, E. shannonea* and *E. tenella*. The Penaeaceae are represented by *Brachysiphon acutus, Penaea mucronata* (also common in other fynbos types on acid soils) and *Saltera sarcocolla*; the Grubbiaceae by *Grubbia rosmarinifolia*; and the Retziaceae by the rare *Retzia capensis*. The broad-leaved sedge, *Tetraria thermalis*, is a common component of ericaceous fynbos. The only proteoid members of this vegetation type are scattered and stunted individuals of *Leucadendron xanthoconus* and *Leucospermum cordifolium*.

Dune asteraceous fynbos

Asteraceous fynbos, which is almost entirely dominated by ericoid shrubs (other than *Erica* species), grows in some of the drier habitats of the Southern Overberg, where soil moisture is too low to support proteoids. Such habitats are found on the well drained coastal dune sands. Like the limestone soils, these dune sands are also calcareous and consequently dune asteraceous fynbos and limestone proteoid fynbos share many species.

Ericoids commonly found include *Acmadenia obtusata, Agathosma collina, Metalasia muricata, Muraltia satureoides* and *Passerina ericoides*. Other prominent shrubs on the coastal dunes are *Helichrysum dasyanthemum, Otholobium fruti-*

The rare and threatened Leucadendron elimense *subsp.* elimense *is a characteristic plant of Elim asteraceous fynbos*

cans, Pelargonium betulinum and *Salvia africana-lutea.* Thicket shrubs, especially *Rhus glauca, Euclea racemosa* and *Olea exasperata,* occur throughout dune asteraceous fynbos. Restioids are not common: *Ischyrolepis eleocharis* and *Chondropetalum microcarpum* are the principal species. Geophytes are plentiful, especially after fire when *Haemanthus sanguineus, Moraea fugax* and *Satyrium carneum* brighten the charred landscape.

Elim asteraceous fynbos

This vegetation type is associated with the shallow, gravelly soils derived from patches of silcretes and ferricretes that still cap much of the Bokkeveld shales in the Southern Overberg. These soils are baked dry in summer, and too shallow to store much moisture after winter rains. This habitat is ideal for the relatively shallow-rooted ericoid shrubs that are prevalent in this vegetation type.

Elim asteraceous fynbos is home to many Southern Overberg endemics and, as an entire vegetation type, is restricted to the western part of this region. The most characteristic species are low, ericoid plants and small-leaved members of the Proteaceae, all of which are restricted to this vegetation type. These include *Leucadendron elimense, L. laxum, L. modestum. L. stelligerum, Leucospermum heterophyllum,* and *Protea pudens. Phylica nigrita* and *Thoracosperma puberulum* are just two of the many ericoids that grow in Elim asteraceous fynbos. The open conditions and relatively fertile soils favour geophytes, including characteristic species such as *Gladiolus guthriei*, *Romulea flava* and *Watsonia coccinea*.

Wet restioid fynbos

Much of the low-lying country in the south-western part of the Southern Overberg, centred on the Soetendalsvlei, is inundated during the rainy season. Here shallow sand overlies impermeable layers of rock or hardpan: the result is waterlogged and anoxic conditions during winter, and bone dry conditions during summer. Woody plants do not thrive in these habitats, but restioids do – hence the name: wet (seasonally waterlogged) restioid fynbos.

The characteristic restioid plant, especially on alkaline sands, is the tall thatching reed, *Chondropetalum tectorum*. *Elegia filacea* occurs on the acidic, seasonally waterlogged sands, such as those south-west of the Soetanysberg at Rietfontein. Other common shrubs are *Leucadendron linifolium*, *Cliffortia ferruginea*, *Orphium frutescens* and *Limonium anthericoides*. Some characteristic geophytes are *Gladiolus carneus*, *Micranthus junceus* and *Sparaxis bulbifera*.

Dry restioid fynbos

Well-drained and fine-grained sands in dry areas also support restioid fynbos. Here the shallow-rooted restioids are capable of absorbing all the moisture as it slowly moves down the soil profile, leaving very little for deeper-rooted shrubs. It is also possible that the tall restioids which grow in these habitats make use of fog that condenses on their long stems and drips down to their shallow roots.

Two forms of dry restioid fynbos occur in the Southern Overberg. *Thamnochortus insignis*, widely harvested as a thatching reed, grows on younger sands, often associated with limestone outcrops. This species occurs in almost pure stands with few companion species. *Thamnochortus erectus*, an inferior thatching reed, grows on older, wind-blown sands, which often occur as a band inland of the *T. insignis* veld. Companion species include the rhizomatous grass, *Cynodon dactylon* (kweek), and many thicket shrubs. The transition between these two forms of dry restioid fynbos is best observed on the road between Bredasdorp and Cape Agulhas.

Renosterveld

Renosterveld is a non-fynbos vegetation type associated with Bokkeveld shale-derived soils of the Southern Overberg. Owing to the relatively fertile and clayey nature of these soils, most Renosterveld has been replaced by agricultural crops. Ericoid shrubs and grasses are the predominant plant types in this vegetation.

The characteristic shrub is renosterbos (*Elytropappus rhinocerotis*) although this species also grows in degraded Elim asteraceous fynbos. Other shrubs typ-

Wet restioid fynbos, with Chondropetalum tectorum, *at the Karsriviervlei south of Bredasdorp.*

Harvesting of restios for thatch or 'dekriet' is an important, veld-based industry in the Southern Overberg.

Coastal strand vegetation on sand dunes at De Hoop. Conspicuous species are Ehrharta villosa, Chrysanthemoides monolifera *and* Stoebe plumosa.

ical of Renosterveld include *Athanasia dentata, Pteronia incana* and *Aspalathus* spp. The grass layer is dominated by *Themeda triandra*, especially in well-preserved remnants. Renosterveld has a great diversity of geophytes, including *Gladiolus gracilis, G. liliaceus, Tritonia deusta* and *Watsonia aletroides*.

Forest and thicket

Forest & thicket covers only a small portion of the Southern Overberg and is usually confined to fire-free sites such as rocky kloofs, river valleys and the coastal margin. True forest occurs in a few small patches in the south west (e.g, Grootbos near Stanford) and on the lower, south-facing slopes of the Potberg. The lower, and virtually impenetrable, thicket vegetation is more widespread. Good examples are found on coastal dunes and along the edge of the De Hoop Vlei. Isolated patches of white milkwood (*Sideroxylon inerme*) thicket, sometimes reaching the stature of a low forest, occur on heuweltjies in the low-lying country of the Zoetendals Vallei. These magnificent trees are a characteristic feature of the Southern Overberg.

Common thicket shrubs and trees include *Carissa bispinosa, Cassine maritima, C. peragua, Cussonia thyrsiflora, Maytenus procumbens, Olea capensis, O. exasperata, Polygala myrtifolia, Pterocelastrus tricuspidatus, Sideroxylon inerme, Tarchonanthus camphoratus* and *Zygophyllum morgsana*. Vines and climbers are abundant in

Wind-pruned and salt-stunted rocky shelf vegetation along the Brandfontein coast, west of Cape Agulhas.

the lower thickets which support such species as *Asparagus asparagoides*, *Cineraria geifolia*, *Cynanchum obtusifolium* and *Solanum quadrangulare*.

Coastal strand and rocky shelf vegetation

There is a distinctive coastal vegetation immediately above the high water line. This habitat is subjected to strong and salt-laden onshore winds and, along the soft, sandy coast, substantial sand movement and abrasion. Only a hardy subset of plants can tolerate these harsh conditions.

Two vegetation types occur in the coastal zone: coastal strand on dunes, and rocky shelf on the hard shoreline. Typical coastal strand species include the grasses *Ehrharta villosa* and *Thinopyron distichum*; herbs *Arctotheca populifolia*, *Dasispermum suffruticosum*, *Senecio elegans* and *Silene undulata*; soft shrubs *Hebenstreitia cordata* and *Thesidium fragile*; succulents *Carpobrotus acinaciformis* and *Tetragonia decumbens*; thicket shrubs *Chrysanthemoides monolifera*, *Myrica cordifolia* and *Rhus crenata*; the ericoid *Passerina rigida*; and the geophyte *Trachyandra divaricata*. Many of these species also occur in the mobile dunefields inland of the coastal zone (e.g. west of Koppie Alleen at De Hoop).

In addition to some of the species listed above, rocky shelf vegetation comprises dense swards of the grass *Stenotaphrum secundatum* and open, stunted vegetation characterised by succulents *Drosanthemum intermedium*, *Othonna den-*

Without doubt the Groot Hagelkraal area is the botanical jewel of the Southern Overberg. This remarkable hot-spot is the exclusive home of six species and supports another 21 regional endemics. We owe it to ourselves and future generations that areas such as this are conserved for posterity.

tata, and *Prenia vanrensburgii* and semi-succulent herbs *Osteospermum fruticosum* and *Plantago crassifolia*. *Coleonema album*, an ericoid shrub, is a characteristic species of rocky shorelines battered by strong onshore winds.

Wetlands

Numerous wetlands are scattered across the coastal flats of the Southern Overberg. In the west are the acidic wetlands associated with the Hagelkraal and Ratel rivers. The Zoetendals Vallei area includes many endorrheic (internal drainage) vleis (including Soetendalsvlei, the biggest natural inland water body in South Africa) and the wetlands of the Nuwejaars River. Prior to artificial drainage and siltation this must have been a magnificent wetland system with floodplain grasslands providing ample grazing for many bontebok and hippopotami. In the east are the wetlands of the De Hoop vlei with its rich bird life, now classified as a Ramsar site. These and the Zoetendals Vallei wetlands are alkaline and relatively saline. They have a different plant life from the acidic wetlands to the west.

The acidic wetlands may support dense fynbos, characterised by *Berzelia lanuginosa* and *Psoralea pinnata*, amongst others. *Prionium serratum* is common

The magnificent wetlands of the De Hoop vlei.

Sadly, most of the western part of the Southern Overberg, such as this scene near Baardskeerdersbos, is a messy mosaic of alien plants, agricultural lands and fynbos remnants.

in the stream bed vegetation. Wetland geophytes include *Gladiolus tristis*, *Spiloxene aquatica* and *Watsonia meriana*. On the shores of the alkaline vleis are reeds and sedges such as *Phragmites australis* and *Juncus krausii*. The saline flats east of the Heuningrug and elsewhere are dominated by *Sarcocornia littorea*. True aquatic plants include *Aponogeton distachyos*, *Limosella grandiflora* and *Nymphaea capensis*.

Alien vegetation
Alien vegetation is becoming an increasingly common feature of the Southern Overberg. The principal alien invasives are wattles, eucalypts and hakeas from Australia and pines from the Mediterranean Basin and California. These areas all have a climate very similar to parts of the Cape Floristic Region.

Acacia cyclops (rooikrans) forms dense thickets on the alkaline coastal sands as well as on sandy patches in the limestone landscape. *Acacia saligna* (Port Jackson willow) and *A. longifolia* (long-leaved wattle) are the main invaders of acid sands. Fortunately biological control agents are halting the spread of these two species. *Acacia mearnsii* (black wattle) is a rampant invader of streamside habitats where it alters the hydrology and ecology of river systems. In the past decade *Leptospermum laevigatum* (myrtle) has spread rapidly over large areas of the Southern Overberg and may soon become *the* major invasive threat. However, biological control agents have recently been introduced to counter its advance.

Conservation and land use

Like most lowland areas in the Cape Floristic Region, very little of the Southern Overberg is included in a formal reserve network. Indeed, only 12% of the land surface is preserved in this way. Provincial nature reserves, owned and managed by Western Cape Nature Conservation, are De Hoop (32 806 ha), Walker Bay (6 688 ha), De Mond (1 768 ha) and Salmonsdam (838 ha). The Heuningberg nature reserve (900 ha) in the eastern Bredasdorpberge, is managed by the Bredasdorp local council. There are also a number of private nature reserves and natural heritage sites in the western (Agulhas Plain) part of the region.

At present much the Southern Overberg is a patchwork of agricultural lands, alien plant thickets and natural vegetation. For example, on the Agulhas Plain, in an area of 1 540 km^2 between Stanford, Bredasdorp and Cape Agulhas, 22.5% of the area is under cultivation or towns and of the remainder, 15% has been densely invaded by alien trees and shrubs. Almost none of the area is alien-free. The great challenge is to prevent further deterioration of the region's conservation status. In this respect the proposed Agulhas National Park may play an important role. There are drives to market the region as a nature tourism destination. Once the economy of the Southern Overberg is substantially buoyed by its natural assets its conservation status will be strengthened.

Choice and arrangements of plates

There are about 2 500 plant species recorded from the Southern Overberg and 512 are described in this book. Our choice was based on the following criteria:
1. Conspicuous plants typical of major vegetation types, especially those not well covered in other guides (*e.g.* plants of Elim fynbos and limestone proteoid fynbos.
2. Plants likely to be seen by holidaymakers (*e.g.* coastal plants) and ecotourists (*e.g.* plants of De Hoop and the other nature reserves) and roadside plants.

Families of plants are arranged in an order which reflects their relationships with each other. The first set of families (Aponogetonaceae to Orchidaceae) are in the Monocotyledon group, in which all the plants have flower parts arranged in threes or multiples of three. The remaining families (Myricaceae to Asteraceae) are in the Dicotyledon group where flower parts are in fours or fives, or multiples thereof. Within these two main groups the families are arranged in order of increasing specialisation. Thus the Orchidaceae are the most specialised Monocotyledon family, and the Asteraceae the most specialised Dicotyledon family.

Companion books

The following books will be useful for identifying plants not described in this book and are obtainable from the Botanical Society Garden Shop at Kirstenbosch:

Bohnen, P. 1986. *Flowering Plants of the Southern Cape*. The Still Bay Conservation Trust, Still Bay.

Bohnen, P. 1997. *More Flowering Plants of the Southern Cape*. The Still Bay Conservation Trust, Still Bay. (Coverage includes the Riversdale coastal plain to the east of our area, the above are useful for limestone proteoid, neutral sand proteoid, dry restioid and dune asteraceous fynbos types)

Burman, L & Bean, A. 1985. *Hottentots Holland to Hermanus. South African Wild Flower Guide no. 5*. Botanical Society of South Africa, Cape Town. (For plants growing in acid sand proteoid fynbos of the Bredasdorpberge)

Jeppe, B. 1989. *Spring and Winter Flowering Bulbs of the Cape*. Oxford University Press, Cape Town. (Out of print)

Rebelo, T. 1995. *Proteas: A field guide to the Proteas of southern Africa*. Fernwood Press, Cape Town. (Good for the widespread proteas)

Schuman, D. & Kirsten, G. 1992. *Ericas of South Africa*. Fernwood Press, Cape Town. (Essential for the serious erica enthusiast)

Scott, A. 1995. *The Overberg Explorer: A guide for environment-orientated travel in the Cape Overberg*. Cape Overberg Tourism Association. (A useful companion for the ecotourist)

Edmondia sessamoides *at Groot Hagelkraal*

❏ = Plants designated with Red Data Book status
★ = Invasive alien species

Flowering season is indicated in brackets at the end of the text for each species

Description of plants

APONOGETONACEAE

Aponogeton distachyos waterblommetjie, wateruintjie
An aquatic plant with a tuberous root and floating leaves (60-200 mm long) borne on long stalks. The forked, sweetly scented, white flowerheads are edible and used to make a Cape stew, "waterblommetjie bredie". It is common in pools and standing water from Nieuwoudtville to Knysna. (July to Dec)

Aponogeton angustifolius
This species is similar to *A. distachyos* but differs in that the flowers are scentless. It occurs in the area and also from Malmesbury to Worcester. (June to Sept)

POACEAE

The grass family comprises herbaceous, often tufted, plants that have hollow stems and leaves that encircle the stem with a split sheath. The flowerheads consist of spikelets, an arrangement of dry bracts around the stamens and stigmas. They have a world-wide distribution.

Cynodon dactylon (fyn)kweek
A perennial grass up to 350 mm high with a characteristic 4-5 pronged whorl of flowering branchlets at the ends of the stems. It occurs along road sides and in most overgrazed areas. The rhizome extends to 1 m down, making it difficult to eradicate. Found world-wide in subtropical areas and is often considered a weed. It occurs throughout most areas of southern Africa, and is valued as it binds the soil and provides some grazing. The roots have medicinal properties. (Sept to May)

Ehrharta villosa var. **maxima**
A robust perennial grass up to 1,5 m, with leaf blades (15-130 mm long) that are rolled and often drop off. The flowering spikelets (12-18 mm long) are straw and purple coloured. It forms large stands on coastal dunes, and is endemic to the southern African coast line from Saldanha Bay to Port Elizabeth. (Sept to March)

Aponogeton angustifolius

Aponogeton distachyos

Cynodon dactylon

Ehrharta villosa var. *maxima*

Hyparrhenia hirta   bosluisgras
A tufted perennial up to 800 mm high with harsh, narrow leaves. Flower spikelets are yellow-green to violet, with white hairs. It grows in stony soils where it forms extensive stands along roadsides in the area. It occurs in most parts of southern Africa, as well as elsewhere in Africa and the Mediterranean Basin. (Sept to June)

Merxmuellera cincta
A tall, densely tufted, perennial up to 2 m high with long leaf blades (1 m long). The cream flower spikelets (*ca.* 14 mm long) are arranged in dense clusters (200-400 mm long). It is locally common in seep areas and along river banks. This species is endemic to fynbos areas. (Sept to Feb)

Merxmuellera stricta
A tufted perennial up to 800 mm, with narrow leaves (*ca.* 450 mm long; 0,5 mm wide) and straw and purple spikelets (23 mm long) that are arranged in loose clusters. It occurs on flats and slopes in the area and is distributed between Nieuwoudtville and George, as well as in the karoo mountains and Drakensberg. (Aug to Mar)

Pentaschistis eriostoma   olifantsgras
A tufted perennial up to 900 mm, with leaf blades up to 400 mm long and straw- and cream-coloured spikelets (*ca.* 10 mm long). The basal part of the leaf sheath usually has a dense, woolly covering. It occurs in most drier fynbos areas, and is also found in the mountainous areas of succulent karoo. (Sept to Nov)

Hyparrhenia hirta

Pentaschistis eriostoma

Merxmuellera stricta

Merxmuellera cincta

Phragmites australis fluitjiesriet
A tall, perennial, bamboo-like plant up to 4 m high with solitary, robust stems and brown and white spikelets (18 mm long). This cosmopolitan species occurs in most of southern Africa where it forms extensive, dense stands in river beds and wetlands. (Dec to June)

Stenotaphrum secondatum buffelskweek, buffalo grass
A perennial plant with a network of stems that spread vigorously along the ground to form a dense turf (up to 400 mm). It has keeled leaf blades that are folded and flattened. This is a hardy plant that is able to withstand the saline conditions of beaches and marshes, and is common along most of the southern African coastline. It is used for pasture, and also forms a resilient, water-wise lawn. (Oct to May)

Pseudopentameris macrantha
A tufted perennial up to 1,2 m tall. Leaves about 500 mm long and about 4 mm wide. The spikelets (*ca*. 45 mm long) are green, purple or straw-coloured. This grass occurs in localised stands in rocky or sandy fynbos-clad slopes between the Cape Peninsula and Bredasdorp. (Aug to Dec)

Stenotaphrum secundatum

Pseudopentameris macrantha

Phragmites australis

Thinopyrum distichum sea wheat
This is a hardy and robust perennial plant reaching 600 mm in height. It has thick, creeping stems that are profusely rooted at the nodes. The spikelets (up to 40 mm long) are hard and smooth, and are arranged alternatively along the stem, thus resembling wheat. This grass is endemic to coastal areas in the southern Cape where it is common on shifting sand dunes. It is able to withstand the sea salt and spray and wetting by spring tides. This efficient sand binder is often used in reclamation work. (Oct to Jan)

Themeda triandra rooigras
A perennial grass up to 1,5 m high with compressed leaf sheaths. The spikelets have long awns and occur in drooping, triangular, reddish clusters. It grows in renoster veldlands and grassy fynbos in the area and is widespread in sub-Saharan Africa. Rooigras includes many varieties most of which provide excellent grazing. (Sept to June)

CYPERACEAE

This sedge family differs from the Poaceae in that the leaf blades encircle the stems, but with no slit. The stems are solid.

Tetraria thermalis bergpalmiet
A tussocky sedge up to 2 m tall with tough, long leaves (*ca.* 1 m long; 200 mm wide) that have sharp, cutting margins. The 3-sided flowering stems stand tall above the tufts of leaves and bear clusters of brown spikelets (flowers). It grows on flats and slopes in relatively moist areas from the Cape Peninsula to Riversdale. (June to Oct)

Tetraria bromoides bergpalmiet
A tall, tufted plant up to 1 m high with grass-like leaves 400-500 mm long. The flower spikelets are pinkish. It is common on the Potberg, and is generally widespread on lower slopes between the Cape Peninsula and Port Elizabeth. (Oct to Feb)

Thinopyrum distichum

Themeda triandra

Tetraria thermalis

Tetraria bromoides

Ficinia filiformis
A densely tufted, slender perennial, up to 200 mm, with narrow, needle-like leaves and brown flowering spikelets (7-10 mm wide). This widely distributed species occurs on flats and slopes from Clanwilliam to the Eastern Cape, as well as in tropical Africa. (Aug to Nov)

Ficinia praemorsa
This tough sedge grows to 500 mm high and has tufts of pointed basal leaves. Compact, rounded clusters of yellow and brown spikelets occur at the top of the tall stems. It is restricted to limestone outcrops in the area. (July to Oct)

Ficinia truncata
A robust, grey-green, tufted perennial about 300 mm tall that resprouts after fire. It has tufts of short, grey-green leaves (*ca.* 60 mm long, 3 mm wide) with papery margins and blunt, square (truncate) tips. The creamy-yellow flowers occur in otherwise chestnut brown spikelets and are clustered at the top of the 200-300 mm high protruding stem. It is confined to limestone hills between Gansbaai and the Gouritz River, and also to similar limestones in the Algoa Basin. (June to Oct)

Hellmuthea membranacea
A reed-like, tufted perennial up to 600 mm tall with a few small (10-50 mm long) leaves at the base of the flower stalk. Dark brown spikelets occur at the tips of the leafless stems. This sedge occurs in dune fynbos from the Cape Peninsula to Knysna. (July to Sept)

ARACEAE

Zantedeschia aethiopica   varkblom, arum lily
A tuberous perennial plant up to 1 m tall with its familiar, spathe-like flower-heads and central, flower-bearing, yellow column. This species is common in damp places from Vanrhynsdorp to Port Elizabeth, and also to the north. It is frequently seen growing along roadsides, and is cultivated worldwide. (June to Feb)

Ficinia filiformis

Ficinia praemorsa

Ficinia truncata   *Hellmuthia membranacea*

Zantedeschia aethiopica

RESTIONACEAE

This family of reed-like plants is always present in fynbos vegetation. Although some of the species are inconspicuous, most are prominent and provide an array of beautiful textures. The leaves comprise short, dry sheaths that are split to the base and spaced at intervals along the stem. Green stems take over the role of photosynthesis. Male and female flowers occur on separate plants – in some cases the respective plants look very different. There are about 400 known species in this family, of which at least 106 occur in the southern Overberg.

Calopsis vimineus
This reed has spreading tussocks up to 1 m in height. The stems are branched and finely ridged, with awns (*ca.* 5 mm long) on the leaf sheaths. It occurs in a wide range of habitats from sandy areas near the coast to inland mountains and is found in fynbos habitats from the Gifberg to Port Elizabeth.

Ceratocaryum argenteum   arrow reed
Female: Plants form large tussocks up to 2,5 m tall with thick (6 mm in diameter), unbranched stems. The flowerheads consist of densely packed spikelets within a large, hard, shiny brown bract (*ca.* 50 mm long). Each spikelet has one flower enclosed by numerous slender papery bracts. It occurs in dense stands in seepages in the area, and further west to Paarl.

Male: Plants are similar in size to the female. The flowerheads have bracts which drop off exposing large numbers of papery flowers.

Chondropetalum microcarpum
This distinctive species has highly-branched stems with spreading leaf bracts. It spreads from an extensive underground rooting system to form "lawns" of plants about 200 mm high. It grows on dunes and exposed limestone plateaux in the area as well as elsewhere along the southern Cape coast.

Chondropetalum tectorum   dekriet
An erect tussock about 1 m tall with unbranched stems. The leaf bracts are deciduous, leaving characteristic dark brown rings at intervals along the green stems where they break off. It occurs in marshes and seeps from Vanrhynsdorp to Port Elizabeth. The plant is harvested for thatch.

Calopsis vimineus

Ceratocaryum argenteum ♀

Ceratocaryum argenteum ♂

Chondropetalum microcarpum

Chondropetalum tectorum

Elegia filacea
This reed grows in erect tussocks up to 500 mm high and has slender, unbranched stems and deciduous leaf bracts. The flowerhead is about 20 mm long and the leaf bract below it is about half as long. Male and female plants are fairly similar. This species often forms extensive stands in sandy and seasonally waterlogged areas, and is found between Vanrhynsdorp and Port Elizabeth.

Elegia muirii
A fountain shaped plant up to 600 mm high with branched stems and persistent leaf bracts. The flowerhead bracts are longer than the leaf bracts along the stem. It occurs in sand or on limestone in coastal areas between Bredasdorp and Riversdale.

Elegia persistens
A fountain-shaped plant up to 1 m high with unbranched stems. The female flowerheads are cylindrical and the flowers are obscured within the brown bracts. Male plants differ in that the bracts are folded outwards, exposing the flowers. This species occurs on rocky mountain slopes between Worcester and Caledon.

Hypodiscus argenteus   foxtail grass
The erect tussocks of this plant grow to about 1 m high. The unbranched stems have fine longitudinal ridges, and transparent, papery leaf bracts. Female spikelets are about 20 mm long, and male spikelets are shorter (*ca.* 5 mm). It grows on dry, well-drained slopes up to 1 200 m and is widespread from the Gifberg to Port Elizabeth.

Hypodiscus willdenowia
A rhizomatous plant (*i.e.* it spreads through a network of underground stems) up to 400 mm in height. The unbranched stems are compressed and have fine longitudinal ridges. It grows on sandy, seasonally wet coastal plateaux in the area, as well as in mountainous areas between Vanrhynsdorp and Port Elizabeth.

 *Elegia filacea*

 *Elegia persistens*

 *Elegia muirii*

Hypodiscus argenteus ♂

Hypodiscus argenteus ♀

 *Hypodiscus willdenowia*

Ischyrolepis capensis
A tangled, fountain-shaped plant up to 500 mm high with branched, warty stems. The floral bracts are long, pointed and curve backwards. It is common in sandy, gravelly or clayey flats or slopes between Clanwilliam and Port Elizabeth.

Ischyrolepis eleocharis
A spreading (rhizomatous) plant growing up to 400 mm tall. It has highly-branched stems, and single spikelets at the ends of stems. This coastal reed occurs on stabilised dunes and limestone hills between Langebaan and Port Elizabeth.

Ischyrolepis leptoclados besemriet
A spreading (rhizomatous) plant with branched stems and single spikelets at the ends of stems. The plants are taller and more finely branched than *I. eleocharis* which it resembles. It is a coastal species found on dunes or limestone outcrops in the area and east to Cape St Francis. The plants are sometimes harvested for brooms.

Mastersiella digitata
A fountain-shaped plant up to 1 m high with branched stems that have tightly rolled leaf bracts. The female spikelets are small and compact, and the male spikelets are cone shaped. It grows on dry, rocky lower slopes in the area, as well as inland to Villiersdorp and west to the Cape Peninsula.

Mastersiella digitata

Ischyrolepis capensis

Ischyrolepis leptoclados

Ischyrolepis eleocharis

Restio multiflorus
A spreading plant up to 1 m high with sparsely branched stems. The flowers protrude beyond the bracts. Scattered plants occur in rocky slopes between Piketberg and Bredasdorp.

Restio triticeus
A fountain-shaped plant up to 800 mm high with stems that have flat-topped whitish tubercles (warts). It is widespread on sandstone-derived soils of lower slopes throughout the fynbos region.

❏ **Staberoha multispicula**
This reed has unbranched stems that form an erect tussock up to 600 mm high. The female flowerhead consists of several spindle-shaped, rigidly erect spikelets, and the male flowerhead consists of up to ten hanging, triangular-shaped spikelets. It grows in sandy or peaty low-lying coastal areas in the Bredasdorp area.

Restio multiflorus

Restio triticeus

Staberoha multispicula ♂

Staberoha multispicula ♀

The genus *Thamnochortus* is recognised by its stiffly erect female spikelets and pendulous, linear male spikelets.

Thamnochortus insignis — dekriet, thatching grass
A striking and elegant plant which grows as a large, fountain-shaped tussock up to 1.5 m. It has unbranched stems with female spikelets up to 25 mm long. The straight, tough stems are ideal for thatching and the species is widely harvested. It occurs on sandy, coastal flats between limestone ridges from Cape Agulhas to Riversdale. Similar-looking *T. erectus* (jakkalsstertriet, elephant reed) is also a tall, tussocky plant about 1,5 m high, but with a rather untidy appearance, and smaller, rounder female spikelets (*ca.* 10 mm long). It has a wider distribution on the coastal forelands between Darling and Knysna.

Thamnochortus fraternus
A fountain-shaped plant up to 1 m high and with erect, narrow, pointed bracts. Female spikelets are about 20 mm long. It occurs on coastal flats and limestone hills in the area as well as west to False Bay. A similar-looking and closely-related species is *T. paniculatus*, endemic to limestone habitats in the Bredasdorp area. These two species are very difficult to distinguish.

❏ Thamnochortus pellucidus
This fountain-shaped, tussocky plant grows to about 600 mm high. Flower spikelets grow on tall, unbranched stems, while the shorter, sterile stems are branched. Several female spikelets (each *ca.* 20 mm long) are clustered together. It is found on sandy, coastal forelands between Gordons Bay and Cape Agulhas.

Thamnochortus fraternus ♀   *Thamnochortus pellucidus* ♀   *Thamnochortus insignis*

Thamnochortus insignis ♀

Thamnochortus fraternus ♂   *Thamnochortus pellucidus* ♂

Thamnochortus insignis ♂

JUNCACEAE

Juncus kraussii
A rigid, tufted perennial up to 1 m high with long, narrow leaves tightly pressed against the stem. Clusters of small, brown flowers occur at the tips of stems. It is found in marshy areas along the coast between the Cape Peninsula and Port Elizabeth, as well as east to Mozambique. It also occurs in Australia and South America. (Sept to Jan)

Prionium serratum   palmiet
This unusual, robust plant grows to 2 m in height and has woody stems that are covered with the dark brown, fibrous remains of old leaves. The stiff leaves are crowded at the ends of branches and have dangerously toothed margins. It forms dense stands in acid-water stream beds and other damp places, and plays an important ecological role in binding river banks. It is widely distributed between the Gifberg and Port Elizabeth, and further to Natal. (Sept to Feb)

COLCHICACEAE (Liliaceae)

Onixotis stricta   vleiblom
A cormous plant 200-500 mm high with three channelled leaves up to 450 mm long. The flowers are crowded at the top of the flowering spike and are pale pink with darker, purple centres. Plants grow in pools and marshes in the area and also from Namaqualand to the Cape Peninsula. (Aug to Oct)

Ornithoglossum viride
A perennial herb with a corm, growing to 300 mm high with concave leaves of varying length that tend to enfold the stem. The flowers face downwards and the petals, which curve backwards, are green with red to brown margins. It occurs between Clanwilliam and Riversdale, as well as in Namaqualand. (June to Oct)

ASPHODELACEAE (Liliaceae)

Bulbine lagopus   geelkatstert
A tuberous perennial up to 400 mm tall with a basal cluster of semi-cylindrical, fleshy leaves (*ca.* 300 mm long) and taller, unbranched, flowering stems. The yellow flowers (*ca.* 7 mm wide) have spreading petals and bearded stamens. It occurs at lower altitudes from Piketberg to Mossel Bay. (July to Sept)

Bulbinella nutans var. nutans   katstert
Up to 1 m tall with erect, fairly narrow, channelled leaves and a tall, dense spike of white to yellow flowers that have unbearded stamens. It occurs on clay soils from Nieuwoudtville to Caledon. (Aug to Sept)

Trachyandra divaricata   hottentotskool
A rhizomatous perennial growing to 800 mm high with basal tufts of narrow fleshy leaves. The much-branched flowerheads bear short-lived, whitish flowers with petals that curve backwards. It is common in the coastal zone between Lamberts Bay and Port Elizabeth. (Aug to Sept)

Juncus kraussii

Prionium serratum

Onixotis stricta

Bulbinella nutans var. *nutans*

Bulbine lagopus

Ornithoglossum viride

Trachyandra divaricata

Aloe arborescens kransaalwyn
Large, branching shrubs up to 2 m in height with terminal rosettes of grey-green to bright green, toothed, recurved and succulent leaves (*ca.* 600 mm long). The conical flowerheads (up to 900 mm long) have one to three branches bearing flowers that are scarlet, orange or pink, and occasionally yellow. It grows on exposed ridges as well as in dense bush in the Caledon area and east along the coast to Mozambique. It also occurs in mountainous habitats of Zimbabwe and Malawi. Used as a hedge plant in the Bredasdorp area. (June to Feb)

Aloe brevifolia duine-aalwyn, kleinaalwyn
This dwarf aloe, about 500 mm high, grows in clusters of densely-leaved, ground-hugging rosettes each about 80 mm (occasionally 300 mm) wide. The broad, triangular leaves (60 mm long, 20 mm wide at the base) are grey-green and the margins and lower surfaces have white spines up to 3 mm long. The flowerheads are unbranched and the tubular flowers (*ca.* 40 mm long) are pale scarlet, occasionally yellow. It grows on clay soils in rocky habitats and is restricted to the Bredasdorp and Caledon areas. (Oct to Nov)

Kniphofia uvaria vuurpyl, red hot poker
A striking, upright plant up to 1 m high with keeled leaves and spikes of red, tubular flowers that turn yellow with age. It grows on the south-facing slopes of Potberg, Soetanysberg and other coastal hills in the area. Widely distributed in wet sites from Clanwilliam to Port Elizabeth. (Oct to Jan)

HYACINTHACEAE (Liliaceae)

Albuca maxima geldbeursie, soldier-in-the-box
A bulbous plant growing to 1,3 m high with several fleshy, channelled, basal leaves. The "cup-and-saucer" flowers are green and white, and dangle from long flower stalks. It occurs in sandy soils and limestone cliffs in the area, as well as along the Cape west coast and Namaqualand. (Sept to Nov)

Aloe arborescens

Aloe brevifolia

Kniphofia uvaria

Albuca maxima

Lachenalia bulbifera rooi viooltjie
A bulbous plant up to 250 mm, with one or two strap-shaped leaves which may be unmarked, or heavily spotted on the upper surface. The flowerhead consists of hanging, tubular flowers which are orange-red with dark green to purple tips. It occurs on sandy soils, dunes and rocky outcrops from the Cape west coast to Mossel Bay. (May to Aug)

Lachenalia contaminata groenviooltjie, wild hyacinth
This species grows up to 200 mm high, and has six to ten narrow, channelled, erect leaves. The white flowers have spreading petals with a dark spot at their tips. It grows in damp areas on sandy flats from Piketberg to Bredasdorp. (Sept to Oct)

Lachenalia rubida rooiviooltjie, sandkalossie
This attractive bulbous plant has one or two lance- or strap-shaped leaves that may be plain or dark-spotted on the upper surface. It is one of the first of the winter bulbs to appear and heralds the onset of the "green season" – flower buds appear soon after the first of the season's rains, before the leaves have fully developed. The hanging, tubular flowers are bright pink to ruby red, with inner petals that protrude and have purple tips with white markings. It occurs in sandy soils of coastal dune fynbos from the west Cape coast to George. (April to July)

Lachenalia muirii
Up to 250 mm high, with one or two linear leaves that wither before the flowers appear. The urn-shaped flowers, having no stalks, are firmly attached to the flowering stem, and are pale blue with dark brown to maroon tips. The inner ring of three petals protrudes conspicuously from the outer three, shorter petals. It is confined to limestone areas from Bredasdorp to Riversdale. (April to July)

Not illustrated here is *L. sargeantii*, a species endemic to the Bredasdorp mountains. About 300 mm tall with two strap-like leaves, it has an unusual and striking flowerhead comprising a bunch of creamy-green, tubular flowers hanging from long, magenta stalks. It usually flowers (Oct to Nov) only after fire.

Lachenalia bulbifera

Lachenalia rubida

Lachenalia muirii

Lachenalia contaminata

Ornithogalum dubium   geel tjinkerintjee, geel viooltjie, yellow chinkerinchee
A bulbous plant up to 500 mm tall with shiny, fairly fleshy leaves (up to 200 mm long) that sometimes have hairy margins. The yellow to deep-orange (occasionally white) flowers are clustered at the top of a slender stem. It is common on renosterveld remnants in the area, and also occurs on flats and slopes from Clanwilliam to Port Elizabeth, and further afield in the Eastern Cape. (Aug to Dec)

Ornithogalum thyrsoides   tjinkerintjee, wit viooltjie, chinkerinchee
This species grows up to 500 mm high and has slender, channelled leaves. The flowers are clustered at the top of the flowering stem and are white with dark, grey-green centres. It is common on flats and lower slopes, usually in damp habitats in the area, and also from the Cape Peninsula to' Namaqualand. Plants are very toxic to stock. The buds were once extensively exported to Europe for the Christmas flower trade. (Oct to Dec)

Massonia pustulata
A low, ground-hugging, bulbous plant with two flat, broad, leathery leaves (*ca.* 15 cm long) that have rough, blistered (pustulate) upper surfaces. The small, creamy-white, pink or yellow flowers are densely clustered between the two leaves. This is a widespread species that occurs in many coastal fynbos areas as well as in Namaqualand. (June to Sept)

Eucomis regia   pineapple flower, pynappelblom
An unusual-looking, bulbous plant 150-200 mm high with up to eight broad leaves flat on the ground, absent at the time of flowering. It has a crown (50-80 mm across) of greenish-cream flowers (the "pineapple") at the top of a stout, purple-spotted stem. It occurs from Nieuwoudtville to Bredasdorp. (Aug to Sept)

Drimia media
Bulbous plant 300-550 mm tall with tufts of somewhat erect, rigid, cylindric leaves. Bulb scales carmine coloured. The flowers are silvery-grey and recurved. It occurs from Saldanha to Knysna in sandy coastal areas. (Jan to Mar)

ASPARAGACEAE (Liliaceae)

Asparagus asparagoides (= *Myrisiphyllum asparagoides*)
krulkransie, Cape smilax
A wiry-stemmed, scrambling plant with stems growing to 3 m as it climbs over other plants. The true leaves are reduced to small, hooked spines and the lance-shaped, apparent leaves are flattened stems (cladodes). It has small, sweetly-scented, white-green flowers and bears red berries. Distribution is from the Gifberg to Port Elizabeth and also in tropical Africa. (June to Sept)

Ornithogalum dubium

Ornithogalum thyrsoides

Massonia pustulata

Eucomis regia

Asparagus asparagoides

Drimia media

HAEMODORACEAE

Dilatris pillansii rooiwortel, bobbejaantou
An erect plant up to 450 mm high with bright red, woody rootstocks and a fan of stiff leaves (*ca.* 300 mm long, 4 mm wide). The velvety, pink-mauve flowers are clustered at the top of a tall, flowering stem and stands are conspicuous after fire. It occurs from Clanwilliam to Bredasdorp. (Aug to Jan)

Wachendorfia paniculata spinnekopblom
A hairy plant about 700 mm high with bright red underground tubers and a fan of pleated leaves (*ca.* 400 mm long). The pale yellow flowers (*ca.* 25 mm long) often have dark markings. It grows in well-drained sands and gravelly soils from Clanwilliam to Port Elizabeth. (Sept to Oct)

Wachendorfia thyrsiflora rooikanol
A robust, plant up to 1,2 m tall with bright red underground tubers and a fan of pleated, hairless leaves (*ca.* 700 mm long, 8 mm wide). The golden-yellow flowers are hairy and are borne on a single, stout stem. Dense stands occur in marshy areas from the Cape Peninsula to Port Elizabeth. (Sept to Dec)

Lanaria lanata kapokblom, perdekapok, Cape edelweiss
An upright plant growing to 800 mm, with numerous stiff, narrow leaves at the base arising from a woody rootstock. The flowering stalk ends in a densely-woolly, white head with hidden, small, mauve flowers. It occurs in the area and east to Port Elizabeth, and blooms profusely after a fire. (Nov to Jan)

Wachendorfia paniculata

Dilatris pillansii

Wachendorfia thyrsiflora

Lanaria lanata

AMARYLLIDACEAE

Amaryllis belladonna. Maartlelie, March lily, belladonna
This striking, bulbous plant grows up to 900 mm, with six to nine erect, strap-shaped, channelled leaves that are absent at the time of flowering. The large, bell-shaped flowers are light to deep pink and are strongly scented. Flowering is spectacular after a fire, at a time when there is not much else in bloom. It is found on lower slopes from Clanwilliam to George. (Feb to April)

Boophane disticha gifbol, seeroogblom
A bulbous plant up to 250 mm tall with a fan of grey leaves that have frilly margins. Leaves are not present at the time of flowering. The pink flowerheads are borne on a compressed stalk. It occurs in the area and east to Port Elizabeth, as well as in subtropical Africa. (Sept to Feb)

Brunsvigia orientalis (konings) kandelaar, candelabra flower
A bulbous plant up to 500 mm high. The two to six tongue-shaped, closely-ribbed leaves are spread flat on the ground and are absent at flowering. The deep pink to bright red flowers are turned upwards and occur in large, round heads. The dry. fruiting heads are wind-blown. This species grows on sandy coastal flats from Saldanha to Knysna. (Feb to April)

Nerine humilis berglelie
A bulbous plant about 350 mm high with basal leaves that are present at the time of flowering. The pink to mauve flowers have wavy margins to the petals and the stamens and styles curve to one side. It grows on sandy flats, slopes and limestone areas from Clanwilliam to Riversdale. (April to May)

Amaryllis belladonna

Brunsvigia orientalis

Boophone disticha

Nerine humilis

❏ **Cyrtanthus guthrieae**   Bredasdorp lily/lelie
This striking fire-lily is a bulbous plant growing to 120 mm in height and has solitary or paired bright red flowers. It flowers after fire on lower sandstone slopes and is endemic to the hills west of Bredasdorp. (Mar to April)

❏ **Cyrtanthus leucanthus**   witbergpypie
A fire-lily with heads of one to three creamy-white flowers (40 mm long) on a plant that grows to 250 mm tall. The narrow leaves (*ca.* 150 mm long; 1 mm wide) grow after flowering has occurred. It is endemic to the Bredasdorp area and occurs on sandy flats and limestone-derived soils. (Jan to Mar)

❏ **Cyrtanthus carneus**   sandlelie, fire lily
A spectacular, bulbous plant up to 1 m tall with long strap-shaped leaves (usually absent at flowering) and large pink to scarlet flowers (up to 75 mm long). It flowers only after fire, and grows in sandy habitats in the area and towards Caledon and Swellendam. (Jan to Feb)

Haemanthus sanguineus   April fool, veldskoenblaar, misryer
A bulbous plant about 300 mm high with two to three large, rounded leaves flat on the ground (like the soles of a shoe) that appear after flowering. The stout, red, flowering stem, often compressed and furrowed, bears the "paint-brush" flowerhead. This consists of a cup of about eight red bracts with numerous pink to red flowers crowded in the centre. It is found on clay or sandy soils from Clanwilliam to Port Elizabeth. (Feb to Mar)

H. coccineus (bloedblom), which also occurs in the area in coastal scrub and lower slopes, differs in that it has two to three tongue-shaped, suberect leaves and a spotted flowering stem.

Cyrtanthus guthrieae

Cyrtanthus leucanthus

Cyrtanthus carneus

Haemanthus sanguineus

HYPOXIDACEAE

Empodium gloriosum   ploegtyd blommetjie
A cormous plant up to 300 mm high with longitudinally pleated basal leaves that appear towards the end of the flowering season. The star-shaped yellow flowers have a long, narrow tube connecting to an ovary that is at ground level. It grows on flats and upper slopes from Clanwilliam to Bredasdorp, in the Little Karoo and also in Natal. (April to June)

Spiloxene aquatica   watersterretjie, vlei-uintjie
A cormous plant with long, erect, linear leaves. The white, star-shaped flowers are loosely clustered at the end of the stem. It grows in pools and marshes from Clanwilliam to Port Elizabeth, as well as in Namaqualand. (June to Oct)

Spiloxene capensis   goue/geelsterretjie
This showy, cormous plant, up to 300 mm high, has stiff leaves that have fine-toothed margins. The large, star-shaped flowers occur singly, and may be white or yellow, sometimes with a dark, iridescent centre. It is found in damp areas from Clanwilliam to Port Elizabeth. (Aug to Oct)

Spiloxene flaccida   sterretjie
This cormous plant, up to 250 mm high, is almost grass-like with narrow, pointed leaves (*ca.* 2 mm wide). The yellow flowers have green undersides and occur in pairs on thin, soft (flaccid) stems. They open in the afternoon. It occurs on damp sandy and gravelly flats and slopes from Clanwilliam to Riversdale, and also in the Little Karoo. (July to Nov)

TECOPHILACEAE

Cyanella lutea   five-fingers, lady's hand
A cormous plant about 250 mm tall with four to six wavy leaves (*ca.* 150 mm long) arranged in a basal rosette. The branched flowerheads bear many, distinctly veined, yellow flowers. The stamens bend downwards and look like fingers on a hand. This widespread plant occurs on sandy flats and slopes on the southern Cape coast as well as in Namaqualand and the Great Karoo. (Sept to Oct)

Empodium gloriosum

Spiloxene aquatica

Spiloxene capensis

Spiloxene flaccida

Cyanella lutea

IRIDACEAE

Aristea glauca
A perennial plant up to 250 mm high with a rhizome. It has winged stems, tufts of tapering, firm-textured leaves (50-90 mm long), and clusters of deep blue flowers. It occurs on coastal sands and lower slopes in the area and elsewhere to the Cape Peninsula and Riversdale. (July to Dec)

Aristea oligocephala
This rhizomatous species grows to 250 mm high and has fairly stiff, cylindrical leaves. Intense blue flowers emerge from translucent bracts. It occurs on flats and slopes in this area as well as at Ceres and Stellenbosch. (Dec to April)

Babiana montana   bobbejaantjie
The name is derived from *bobbejaan* (baboon) because these animals feed on the corms of these plants. This species is a dwarf, bulbous plant up to 70 mm high with hairy, pleated leaves. The asymmetrically shaped flowers are mauve with yellow and purple markings and one of the three stamens is longer than the other two. It is restricted to the Klein River mountains and the Bredasdorp area where it grows on flats and slopes. (June to Aug)

Babiana patersoniae
A small, cormous plant about 100 mm tall with a fan of hairy, pleated leaves. The strongly scented flowers are whitish to pale blue or purple with yellow markings. It occurs on both rocky and deep, sandy soils from Caledon to Port Elizabeth. (Aug to Oct)

Babiana patula
This dainty plant is about 80 mm tall with pleated, hairy leaves and fragrant blue to mauve or yellow flowers. It occurs in the area and also in the Little Karoo. (Aug to Sept)

Aristea glauca

Aristea oligocephela

Babiana montana

Babiana patersoniae

Babiana patula

Bobartia longicyma subsp. **magna** (besem)biesie
A robust, tufted plant up to 1,8 m high with cylindrical, erect leaves as high as the stems. On first sight it resembles a member of the Restionaceae. The rich yellow flowers occur in a cluster near the tip of the flowering stem. It occurs on sandy flats and lower slopes in the area, as well as on the Cape Flats, and is most commonly seen in flower after fire. Used for making baskets in the Elim area and also to make sturdy brooms for cleaning out stables. (Aug to Nov)

Chasmanthe aethiopica suurkanol(pypie)
A cormous plant up to 650 mm tall with sword-shaped, but softish leaves (*ca.* 300 mm long). The orange, tubular flowers are curved with the uppermost, long lobes forming a covering hood over the arched stamens and style. Flowers are positioned on the upper side of the arching flowering stem. It occurs in dune thicket in the area, and elsewhere in dampish, sheltered places in coastal areas between the Cape Peninsula and Transkei. (April to July)

Ferraria crispa spinnekopblom
This leafy and highly-branched cormous plant is about 800 mm high. The flowers are speckled brown with crisped, curly petal margins and fringed styles. They have an unpleasant, foetid odour which attracts flies for pollination. It grows on sandy, coastal soils between Lamberts Bay and Agulhas as well as in the Little Karoo. (July to Oct)

❏ **Freesia elimensis**
A cormous plant up to 150 mm high with erect, sword-shaped leaves. The scented, tubular flowers are white, flushed with mauve on the reverse, and two of the lateral petals have yellow markings. Flowers are borne on the upward-facing side of the arched stem. This species grows only in limestone fynbos, and is restricted to the Bredasdorp area. (May to June)

Freesia leichtlinii kammetjie, freesia
This species grows up to 200 mm high and has upright leaves. The flowers are cream and yellow, often with mauve markings. It occurs on coastal sands from Cape Agulhas to Mossel Bay. (Aug to Sept)

Ferraria crispa

Freesia elimensis

Freesia leichtlinii

Bobartia longizyma subsp. *magna*

Chasmanthe aethiopica

Geissorhiza heterostyla
A cormous plant up to 450 mm high. The three linear leaves (2-5 mm wide) reach up to about half the length of the plant and have raised, hairy margins that are set at an angle to the blade. The branched stem bears up to seven star-like flowers (*ca.* 2,5 cm wide) that have a wide range of colours – white to yellow, occasionally with pinkish purple on the reverse of the petals, to pale blue or mauve and occasionally with a dark purple throat. The styles vary from being short with long curled branches, to long with short recurved branches. It occurs in sandy areas between Clanwilliam and Port Elizabeth. (Aug to Oct)

Geissorhiza inflexa
A cormous plant up to 300 mm, with three narrow leaves (*ca.* 4 mm wide) that usually do not reach beyond the flowering spike, and with raised, angled margins. The flowers may be pink to purple to red, or white to pale yellow and are usually maroon on the reverse. Flowers close up at night. They occur on clay flats and slopes and are typically seen in Elim fynbos. They also occur between the Cape Peninsula and Piketberg. (Aug to Sept)

Geissorhiza ovata
A small, cormous plant up to 150 mm high with two to three leaves of which the upper one resembles a bract on the upper portion of the stem and the lower oval leaves (*ca.* 30 mm long 12 mm wide) are inclined to the ground. The unbranched stem bears up to six long-tubed flowers that are white to pink above with a maroon reverse. Flowers close up at night. This species is common after fire and, although it usually occurs in montane habitats in the south western Cape, it is found at lower altitudes in the Overberg and is common in Elim fynbos habitats. (July to Oct)

The genus *Gladiolus* comprises plants that have a corm and have a spike of one or more showy flowers with a curved tube and uneven petal lobes. This smoking-pipe form has led to the vernacular name "pypie" for many of the species. There are 23 species in this area.

Gladiolus abbreviatus (= *Homoglossum abbreviatum*)
Up to 650 mm tall with linear leaves. The reddish-orange flowers have very short lower petals that tend to be green to blackish. Plants occur in clay soils in the area and also near Worcester and eastwards to Riversdale. (June to Sept)

Gladiolus brevifolius   rooipypie
A plant up to 500 mm, with two to three linear, hairy leaves that are not seen at the time of flowering. The three to ten flowers (*ca.* 40 mm long) are pink to grey-mauve and the lower lobes have pink or yellow markings. It occurs in sandy soils between Clanwilliam and Bredasdorp. (Mar to May)

 *Geissorhiza heterostyla*  *Geissorhiza ovata*  *Geissorhiza inflexa*

Gladiolus abbreviatus

Gladiolus brevifolius

Gladiolus bullatus   berg (mountain) bluebell
A slender plant up to 700 mm high with up to four linear leaves ensheathing the stem. There are one or two nodding, bell-shaped flowers (*ca.* 50 mm long) that are blue to mauve with a band of yellow on the lower lobes. It occurs on sandy slopes in the area and east to Humansdorp. This species is extensively picked for spring flower shows. A warning to flower pickers – if plucked without leaving a leaf, the bulb can rot. (Aug to Oct)

Gladiolus carinatus   blou afrikaner
A slender plant up to 1 m tall with three linear leaves that have a prominent midrib. There are numerous (up to 15) fragrant flowers that are blue to violet, occasionally yellow or pinkish-red and have yellow markings and dark streaks on the three lower lobes. It grows on sandy flats and slopes from Nieuwoudtville to Mossel Bay. (Aug to Sept)

Gladiolus carneus   bergpypie, large painted lady
This plant grows up to 1 m tall, and has one to five distinctly ribbed leaves (*ca.* 600 mm long, 20 mm wide). There are about three white to pink flowers with a deep pink throat and markings on the lower lobes. It occurs locally in sandy soils, often in marshy restioid fynbos areas and also in the Cape Peninsula, Malmesbury and Ceres. (Oct to Nov)

Gladiolus cunonius (= *Anomalesia cunonia*)   lippypie
A plant up to 450 mm high with up to four narrowish leaves (*ca.* 300 mm long). The bright red flowers have a hooded upper petal flagged by two ear-like side petals and three small lower petals. It is found in sandy coastal areas from the Cape Peninsula to Knysna. (Sept to Oct)

Gladiolus debilis   painted lady, kalkoentjie
A delicate-looking plant up to 650 mm tall with one to four linear leaves (2 mm wide) that have raised margins. The flowers (*ca.* 400 mm long) are white to pink with deep pink-red markings on the lower petals. It occurs on acid and limestone sandy flats and slopes in the area as well as west to the Cape Peninsula. (Sept to Oct)

Gladiolus bullatus

Gladiolus carinatus

Gladiolus carneus

Gladiolus cunonius

Gladiolus debilis

Gladiolus gracilis
sandpypie, reed bells, bloupypie

A slender plant with about four narrow leaves (*ca.* 600 mm long) that are often rolled inwards giving a cylindrical appearance. The fragrant flowers (*ca.* 30 mm long) are pale blue, occasionally pale pink to mauve, with purple and yellow streaks on the lower petals. This species occurs on sandy flats and slopes, as well as on gravelly and clay-rich renosterveld areas. Its distribution extends westwards to the Cape Peninsula and Malmesbury. (May to Sept)

Gladiolus guthriei
A plant up to 700 mm high with one to three hairy, narrow, tapering leaves that are not seen at the time of flowering. The sweetly-scented flowers are pink to brick-red with gold or black frilly edges to the petals. The top petal is hooded and the lower three curl down. It is confined to the Bredasdorp area and grows in gravelly Elim fynbos as well as in sandy, coastal areas. (May to July)

Gladiolus liliaceus
groot bruin afrikaner

A plant up to 800 mm high with two to four narrow, linear leaves that have thickened margins. This extraordinary flower changes colour from rusty-red during the day to bluish-mauve at night, at which time it also becomes strongly fragrant. It is distributed between Clanwilliam and Port Elizabeth and occurs in clay and loam-rich soils in the Agulhas area. (Aug to Nov)

Gladiolus maculatus
klein bruin afrikaner, kaneelpypie

Up to 700 mm high, with three to four linear leaves that are not seen at the time of flowering. The strongly-scented flowers vary from cream-yellow to pink with dark brownish spots. The petal margins are wavy and the tips tend to curl backwards. It occurs on lower slopes between the Cape Peninsula and Grahamstown. (Mar to April)

Gladiolus meridionalis
This striking and rare plant, previously considered to be a subspecies of *Gladiolus maculatus*, is about 400 mm high and has two to four pink-red, unscented flowers. It is found only in a few sandy areas between Pearly Beach and Elim. (May to July)

Gladiolus gracilis   *Gladiolus guthriei*   *Gladiolus liliaceus*

Gladiolus meridionalis   *Gladiolus maculatus*

Gladiolus miniatus (= *G. floribundus* subsp. *miniatus*)
This species grows up to 400 mm high and has a fan of four basal leaves. The bottom of the stem is speckled. The long and wide-necked flowers are cream to salmon-pink with dark stripes down the petals, and are arranged to one side of a zig-zag stem. It grows on limestone. (Oct to Nov)

Gladiolus pillansii
A plant up to 600 mm high with cylindrical leaves that grow after the flowering season. The small, fragrant flowers are blue-grey to pink with maroon and yellow markings on the lower petals. It occurs in flat, sandy areas between Clanwilliam and Stanford. (Mar to May)

Gladiolus punctulatus
A plant up to 600 mm high with three hairy, linear leaves. The flowers vary in colour from pink to mauve, with dark spots and streaks. It occurs on sandy flats and slopes between Piketberg and George. (June to Oct)

Gladiolus rogersii   dune bluebell
A slender plant up to 600 mm high with three to five narrow, linear leaves. It bears 5-14 flowers that are pale to deep blue, or pink-mauve with distinctive yellow and dark purple markings on the three lower petals. It occurs on sandy and gravelly soils as well as on limestone between Bredasdorp and Humansdorp and in the Little Karoo. (July to Nov)

❏ **Gladiolus stefaniae**
A plant up to 400 mm high with one to three linear leaves that have thickened margins. Flowering occurs after the leaves have died back. Flowers are red with white markings. Rare it is found on mountain slopes in the area and also near Montagu. (March to April)

Gladiolus miniatus

Gladiolus pillansii

Gladiolus rogersii

Gladiolus punctulatus

Gladiolus stefaniae

Gladiolus tenellus
geelkalkoentjie, botterpypie, vleipypie

A slender plant up to 400 mm, with three narrow, cylindrical leaves that are taller than the inflorescence. The evening-scented flowers are cream to yellow and may be flushed with pink or red. This species occurs in marshy areas on flats and lower mountain slopes between Piketberg and Bredasdorp. (July to Oct)

Gladiolus teretifolius (= *Homoglossum muirii*)

A plant about 600 mm high with narrow, linear leaves that have thickened margins and midribs and are slightly spirally twisted. The red to orange flowers have a long, bent, cylindrical tube that flares into six lobes at the end. It occurs between Stellenbosch and Mossel Bay. (May to Sept)

Gladiolus tristis
aandblom, vlei-aandblom

A tall, stately plant up to 1,5 m tall with two to four narrow, linear leaves (as tall as the inflorescence) that have thickened margins and are spirally twisted in the upper half. The flowers become strongly fragrant in the evenings and are yellow to green, sometimes with dark stripes on the three upper lobes. This species grows in marshy areas and on stream banks in both flats and mountain slopes from Clanwilliam to Port Elizabeth. (Sept to Dec)

Gladiolus vaginatus

This species grows up to 700 mm tall and has two to three cylindrical leaves that are not seen at flowering time. The scented flowers are blue to grey with an orange throat and striped lower lobes. It occurs on flats and lower slopes in the area and east to Knysna. (Feb to April)

Gladiolus variegatus

This species was originally considered to be a more robust variety of *G. debilis* which it resembles, except that it has larger flowers (whitish-pink with deep red-maroon markings), stouter stems and only three leaves. It occurs in limestone habitats in the Gansbaai, Bredasdorp and Agulhas areas. (Sept to Oct)

Gladiolus inflexus

This is a newly described species with blue-mauve flowers that darken towards the edge of the petals and dark spots on the lower petals. It grows in limestone areas in the Bredasdorp area. (June to August)

Gladiolus tenellus

Gladiolus teretifolius

Gladiolus tristis

Gladiolus vaginatus

Gladiolus variegatus

Gladiolus inflexus (*in ed.* Goldblatt and Manning)

Homeria galpinii
tulp

A cormous plant up to 300 mm high with a solitary, short, channelled leaf (*ca.* 150 mm long; 10 mm wide). The yellow flowers have petals that form a narrow cup at the base and flare out horizontally above. It grows on sandy soils in mountainous areas from Piketberg to Bredasdorp. (Mar to Aug)

Homeria bulbillifera
uintjiestulp

A cormous plant up to 500 mm high with one long leaf (*ca.* 7 mm wide). The pale yellow to orange flowers (*ca.* 40 mm long) are borne on drooping stems. At the end of the flowering season clusters of corms (small, bulb-like structures) develop on the stem. This species occurs on coastal sandy soils, limestone areas and in renosterveld in the area as well as from the Cape Peninsula and Port Elizabeth. (July to Oct)

A characteristic of the genus *Moraea* is that the three outer petals differ in size and shape from the three inner ones.

Moraea tripetala
kleinuintjie, perde-uintjie

This species is about 500 mm tall and has one or two linear leaves. The pale blue to purple (occasionally yellow or pink) flowers have three showy petals with light patches of "furry" markings. It occurs on sandy or clay soils and is common in Elim fynbos. It also grows westwards to Nieuwoudtville and eastward to George as well as in the Little Karoo. (Aug to Sept)

Moraea neglecta

This species was only recently named, hence the specific name "neglecta". It is an unbranched plant up to 500 mm high with a single, rigid leaf that is taller than the inflorescence. The stem bearing the flowers is sometimes sticky and the cream to bright yellow flowers that open at midday have characteristic black-dotted nectar guides. Plants grow in sandy flats and slopes in the area, as well as from Nieuwoudtville to Caledon. (Sept to Nov)

Moraea fugax

A plant with one to two channelled, long and drooping leaves. The fragrant, short-lived flowers open at midday and vary in colour from white to yellow to mauve, usually with a yellow nectar guide. The frilly, erect structures forming a ring at the centre of the flower are modified styles that look like petals. This species occurs in dune fynbos areas, as well as in gravelly and rocky habitats from Namaqualand to Caledon area. (Aug to Oct)

Homeria bulbillifera

Homeria galpinii

Moraea fugax

Moraea tripetala

Moraea neglecta

Romulea flava var. **flava**
These cormous plants are about 300 mm high and have a tuft of narrow, basal leaves (1-3 mm wide) that often have wide grooves. The solitary white to creamy yellow and occasionally blue flowers occur singly at the top of a slender stem, and have a short tube with spreading petal lobes. It is common in the gravelly ground of Elim fynbos and also occurs eastwards to Riversdale. (June to Sept)

Romulea rosea var. **reflexa** froetang, knikkertjie
Plants are about 400 mm tall and have tufts of narrow (0.5-2.5 mm wide) leaves. The white to pinkish-lilac or magenta flowers have an orange-yellow cup, and often a violet-blue zone in the throat. The outer petals may be irregularly blotched or have dark veins. It is common in the gravelly sands of Elim fynbos, and is also found eastward to Riversdale. (July to Nov)

Ixia micrandra kalossies, perdebiesie
A cormous plant about 600 mm high with two (rarely three) needle-like (*ca.* 1,5 mm wide) leaves that are about half as tall as the plant. The single stem bears two to six white to pale pink or mauve flowers (*ca.* 20 mm wide) comprising six petals flaring out to form a short basal tube. It occurs on the hills and lower mountain slopes in the area as well as further north to Montagu and Oudtshoorn in the Little Karoo. (July to Aug)

Sparaxis bulbifera fluweeltjie
A cormous plant up to 450 mm high with tapered, linear leaves (*ca.* 300 mm long). The flowerhead is branched and the flowers are white to cream, flushed with purple on the underside. After flowering small corms (bulb-like structures) develop at the base of the leaves. This species is locally common in damp, sandy or clay areas and is found from Darling to Bredasdorp. (Sept to Oct)

Tritonia deusta
A plant up to 250 mm tall with a corm and with leaves that are shorter than the flowering stem. The cup-shaped flowers are orange to red with conspicuous veins and may have dark, raised blotches in the throat. It is found on clay flats and slopes between the Cape Peninsula and Riversdale. (Sept to Oct)

Romulea flava var. *flava*

Romulea rosea var. *reflexa*

Tritonia deusta

Ixia micrandra

Sparaxis bulbifera

Lapeirousia pyramidalis
A cormous plant up to 100 mm high with narrow, tapering leaves. The flower-head has a pyramid shape at the bud stage, later opening to a cluster of fragrant, long-tubed flowers that are white to pale blue or pink, with occasional dark markings. This species grows on lower clay slopes and renosterveld areas in the southern Cape as well as in Namaqualand. (July to Sept)

Micranthus junceus   vleiblommetjie
A cormous plant up to 450 mm high with several cylindrical leaves. It has numerous, small, blue flowers that are closely set in flat, two-ranked spikes. Each flower has a short tube that ends in slightly wavy petal lobes. It grows in wet, sandy sites between the Cape Peninsula and Clanwilliam. (Nov to Feb)

Tritoniopsis dodii
A cormous plant up to 500 mm high. The linear leaves have two to three longitudinal veins, and the dull, pink flowers have petal lobes with a dark streak and that are longer than the tube. It occurs on lower mountain slopes between the Cape Peninsula and Bredasdorp where it is locally common on the Potberg. (Feb to April)

Tritoniopsis apiculata
This species grows up to 400 mm high. The sword-shaped leaves have three to five longitudinal veins and are not seen at the time of flowering. The long-tubed, pink flowers have petals with a dark streak. It grows on the stony, lower slopes of the Potberg and eastwards to Mossel Bay. (Mar to May)

Tritoniopsis antholyza (= *Anapalina nervosa*)
A cormous plant growing to as high as 900 mm. The sword-shaped basal leaves have three to six veins, and are withered at the time of flowering. It has a spike of pink to red flowers that have a long, curved tube ending in relatively short petals of which the uppermost one is the longest. This is a plant of rocky mountain slopes such as the Potberg and occurs from the Gifberg to Port Elizabeth. (Dec to Mar)

 *Tritoniopsis dodii*

 *Tritoniopsis apiculata*

 *Tritoniopsis antholyza*

Lapeirousia pyramidalis

Micranthus junceus

The genus *Watsonia* comprises cormous plants restricted to southern Africa. The leaves are firm and sword-like and the flowers that occur in a spike-like arrangement along the stem have well-developed tubes ending in six flared, more or less equal, petal lobes.

Watsonia aletroides   rooi(pypie)
Up to 600 mm high with four to five shortish, shiny leaves (5-10 mm wide) that reach about half way up the flowering spike. The single stems (rarely branched) bear up to 20 red-orange, occasionally pink-purple flowers which have characteristic long, pendulous tubes (*ca.* 40 mm long) ending in short, inconspicuous petal lobes. It grows in clay soils in renosterveld from Caledon to Knysna. (Aug to Oct)

Watsonia coccinea   aandpypie
Up to 300 mm, with four to six leaves (2-8 mm wide) that are between half as long and as long as the plant. The single, unbranched stems bear three to six orange, scarlet, purple or pink flowers (*ca.* 60 mm long) that sometimes have a darker line down the middle of the petals. Flowering is common for about two years after fire. It grows on flat, sandy areas, usually in seeps, between Malmesbury and the southern Overberg. (Aug to Nov).

Watsonia meriana   lakpypie, suurkanolpypie
This species is very similar to *W. coccinea* but is a much larger plant growing up to 2 m. It has four to six leaves (12-25 mm wide) that are about half as long as the plant. The unbranched stems have about 25 bright orange to red, occasionally pink or purple, flowers (*ca.* 70 mm long). It occurs in deep or shallow rocky sands in seasonally moist habitats between southern Namaqualand and Bredasdorp. (Sept to Nov)

Watsonia laccata   waspypie
A plant about 500 mm high with four to five glossy leaves (6-20 mm wide) that are about half as long as the plant. The flower spike bears 8-20 pale pink to purple or light orange flowers (*ca.* 35 mm long). It occurs in seasonally wet, low-lying habitats between Bredasdorp and Knysna. (Aug to Oct)

Watsonia fergusoniae
This plant is up to 800 mm high, and has two to four leaves (*ca.* 5 mm wide) that reach up to the bottom of the flower spike. The unbranched stem has 7-18 orange to scarlet-red flowers (*ca.* 65 mm long). It grows on limestone hills and plateaux along the coast between Agulhas and the Gouritz River mouth. (Oct to Nov).

Watsonia coccinea

Watsonia laccata

Watsonia meriana

Watsonia fergusoniae

Watsonia aletroides

ORCHIDACEAE

Bartholina burmanniana spider orchid, spinnekoporgidee
This dainty and delightful orchid often escapes notice. It has one smooth, heart-shaped leaf lying flat on the ground, and a solitary, pale blue to mauve flower on a 100-200 mm stalk. The much-divided lower lip gives it a spidery appearance. It is occasionally seen on flats and lower slopes from Nieuwoudtville to Humansdorp. (Aug to Oct)

Disa cornuta golden orchid
Robust plants up to 1 m high with overlapping leaves sheathing the length of the stem to below the inflorescence. The flowerhead is 150-400 mm long with numerous white, green and purple flowers. It is often seen flowering after fire, and occurs in well-drained flats and slopes from Clanwilliam to Port Elizabeth and northwards to Zimbabwe. (Sept to Feb)

Disperis capensis moederkappie, granny bonnet
A slender plant up to 350 mm high with two narrow leaves (*ca*. 80 mm long). The flowers are purple to magenta, occasionally white or green, with the upper petals forming a spurred cap and the long, trailing, side petals forming the ribbons of a "granny bonnet". It is occasionally seen on mountain slopes between the Cape Peninsula and Port Elizabeth, and eastwards to the Transkei. (July to Oct)

Herschelianthe purpurascens early blue disa
This orchid grows to 500 mm and has a tuft of narrow, basal leaves. The scented, blue and purple flowers are hooded. Plants are occasionally seen on sandy or gravelly coastal flats between the Cape Peninsula and Agulhas. (Oct to Nov)

Herschelianthe lugens green/bearded disa
This tall orchid grows to 1 m in height and has tufts of slender, basal leaves. The hooded flowers are cream, green and mauve with dark purple markings and the lower lip is highly divided and curly resembling a beard. There are up to nine flowers on a spike. It grows among reeds on coastal flats and is only rarely seen. The distribution is between the Cape Peninsula and Port Elizabeth. (Oct to Jan)

 *Disa cornuta*
 *Disperis capensis*
 *Bartholina burmanniana*
 *Herschelianthe purpurascens*
 *Herschelianthe lugens*

Satyrium coriifolium ewwa-trewwa
Erect plants up to 480 mm, with stout stems and two to four leathery leaves (30-150 mm long) standing free from the ground, and with purple spots at the base. The stem has about 20 bright yellow to orange, spurred flowers that are closely packed against the stem. It occurs in moist, sandy areas of coastal flats and hills, often forming small colonies, from the Western Cape to Grahamstown. (Aug to Oct)

Satyrium carneum (Dryander) rooi-trewwa, pink orchid
This species is similar to *S. coriifolium* but may be taller, up to about 700 mm and the lowest two leaves are pressed flat onto the ground. The flowers are pale pink-rose. It grows on coastal sands and limestone soils from the Cape Peninsula to Riversdale. Indiscriminate flower picking, coastal resort development and alien plant invasions have reduced the numbers of this plant. (Sept to Nov)

Bonatea speciosa Oktoberlelie, green wood orchid
A robust, leafy, almost succulent plant up to 1 m tall with large, broad leaves (*ca.* 150 mm long) sheathing the stem. The green and white flowers have a deeply-lobed lower lip and are clustered at the top of the flowering stem. It grows in damp, sandy areas in coastal and lower mountain slopes between Malmesbury and Port Elizabeth, as well as in the Transvaal. (Oct to Nov)

Monadenia bracteata
An orchid about 500 mm tall with long, tapering, channelled leaves (*ca.* 200 mm long) clustered at the base of the plant and shorter ones ensheathing the flowering stem. The numerous, small, green and reddish-brown flowers have a spurred, hooded cap. This species is often seen in disturbed areas on flats and slopes between Clanwilliam and Port Elizabeth. (Sept to Nov)

Pterygodium catholicum oumakappie, cowled friar
An orchid up to 300 mm high with oval leaves (*ca.* 100 mm long) that encircle the stem. The cream to sulphur yellow, hooded flowers turn reddish and they have an acrid scent. Plants grow in moist, sheltered, clayey areas where they may be locally abundant. This species occurs from Nieuwoudtville to Port Elizabeth. (Aug to Nov)

Satyrium carneum

Satyrium coriifolium

Monadenia bracteata

Bonatea speciosa

Pterygodium catholicum

MYRICACEAE

Myrica cordifolia wasbessie, waxberry, glashout
A sprawling to erect shrub up to 1,5 m high. The small, heart-shaped leaves have serrated margins and are closely packed on the stem. Tiny, inconspicuous male and female flowers occur on different plants. The prolific, dark, round fruits are covered with wax which was extracted and used as a polish by early colonists. Bark was used for tanning skins and this useful plant is also effective as a sand-binder on coastal dunes where plants grow in large colonies. It occurs in sandy areas between the Cape Peninsula and Port Elizabeth. (All year)

Myrica quercifolia maagpynbossie
An erect shrub with a fire-resistant rootstock growing up to 500 mm. It has lobed leaves (ca. 20 mm long) and yellowish flowers arranged on hanging catkins. This species occurs on sandy flats and slopes between Malmesbury and Uitenhage, as well as in Namaqualand and Transkei. It is reputed to cure stomach pains. (July to Sept)

PROTEACEAE

This family of perennial shrubs or trees is largely restricted to the southern hemisphere, with Australia and South Africa being the main centres of distribution. Of the 329 South African species there are 77 in the southern Overberg. They range from low, ground-hugging shrubs (*Protea denticulata*) to small trees (*Protea nitida*). The small flowers (florets) are grouped together in large, often showy, flowerheads. Many are valuable components of the wildflower trade and this is one of the best studied families in fynbos.

*Hakea gibbosa rock hakea
A declared noxious weed, this small, prickly tree grows to 3 m, and has long, hairy, needle-like leaves. The flowerheads comprise small, cream florets. It was introduced from Australia and here forms dense, impenetrable thickets in sandstone-derived soils in the area, completely suppressing the natural vegetation. Every effort should be made to eradicate this alien weed. (June to July)

Aulax umbellata sekelbos, Christmasblom, broad-leaf featherbush
An erect, single-stemmed shrub up to 2.5 m tall with leaves that are almost linear (20-110 mm long, 2-15 mm wide). The separate male and female plants are shown together. Female flowerheads are yellow, forming cup-shaped woody structures in which seeds are retained for several years. Male flowerheads (*ca.* 40 mm long) occur in clusters of 5-15. This species grows in acid sands of coastal areas, often forming extensive stands that are showy in mid-summer. (Nov to Feb)

Myrica cordifolia

Myrica quercifolia

Hakea gibbosa ♂

Hakea gibbosa ♀

Aulax umbellata ♀ and ♂

In the genus *Leucadendron* there are separate male and female shrubs. The female plants bear characteristic cones (tolle).

Leucadendron coniferum — geelbos, rooitolbos, dune conebush

Female: An erect, single-stemmed shrub up to 4 m. The slightly twisted leaves are oblong with a sharp, fine tip and are hairless when mature (*ca.* 8 mm long, 9 mm wide). The mature red cones (45 mm long, 30 mm wide) later become green and are retained on the plant for several years. Involucral leaves are yellow. It occurs together with *Protea susannae* in neutral sand proteoid fynbos in the area. It also grows in the Cape Peninsula and on the coastal flats near Betty's Bay. Enormous quantities of cones are harvested in the wildflower trade. (Aug-Sept)

Male: Similar to female, but with slightly narrower leaves (7 mm) and bearing numerous small flowerheads (18 mm wide).

Leucadendron meridianum — geelbos, limestone conebush

Female: A single-stemmed shrub to 2 m. The leaves are twisted at the base, have silver hairs and a short, red, recurved fine tip (40 mm long, 7 mm wide). Involucral leaves are yellow and are longer than the foliage leaves. The rounded cones are covered with short, silver hairs (16 mm long, 12 mm wide) that are retained on the plant for several years. It is endemic to limestone areas (limestone proteoid fynbos) from Gansbaai to the Gouritz River Mouth and inland as far as Bredasdorp, often forming dense stands. (July)

Male: Similar to the female. The small, yellow flowerheads have a lemon scent.

Leucadendron modestum — skilpadbossie, rough-leaf conebush

Female: A single-stemmed shrub, with red branches, growing up to 600 mm. The oblong leaves feel gritty and have a blunt, red tip (25 mm long, 5 mm wide). Involucral leaves are longer than the foliage leaves. The cones (23 mm long, 15 mm wide) have long hairs, and are kept on the plant for several years. Dense, isolated stands occur on gravelly and clayey soils of the Elim flats (Elim asteraceous fynbos), and also in the Bot River Valley. (Aug)

Male: Similar to the female but with slightly narrower leaves (18 mm). The flowerheads have an unpleasant odour.

Leucadendron coniferum ♂

Leucadendron meridianum ♀

Leucadendron modestum ♀

Leucadendron meridianum ♂

Leucadendron modestum ♂

Leucadendron coniferum ♀

❑ **Leucadendron elimense** subsp. **elimense**

bergkatjiepiering, Elim conebush

Female: An erect, sparsely branched, single-stemmed shrub up to 1,5 m. The hairless, elliptic leaves have a recurved, red, fine tip and vary considerably in size (14-57 mm long, 7-21 mm wide). Involucral leaves are yellow. The flowerheads have a pungent odour and form round cones (35 mm wide) from which seeds are released each year. This species is restricted to the shallow, gravelly soils on shale (Elim asteraceous fynbos) from Gansbaai to Bredasdorp, a habitat that is used for agriculture. Thus, there are only about 2 000 plants remaining in a few isolated, often roadside, populations. (July to Sept)

Male: Similar to the female, but with smaller leaves (13-49 mm long, 5-19 mm wide) and striking, large flowerheads (*ca.* 46 mm wide).

❑ **Leucadendron laxum**

vleirosie, Bredasdorp conebush

Female: An erect, single-stemmed shrub up to 1,5 m, with needle-like leaves (*ca.* 18 mm long, 1.5 mm wide). The attractive, yellow (becoming reddish), oval cones (20 mm long, 14 mm wide) have hairy bracts with elevated tips, thus resembling opening rose-buds. Seeds are shed each year. Plants occur in level, damp ground at the bottom of valleys from Hermanus to the Bredasdorp area where they occur in Elim asteraceous fynbos. Although they may grow in dense stands, this species is under threat as their habitat is often drained for farming. Only about 5 000 plants are believed to remain. (Sept to Oct)

Male: Similar to female with slightly smaller leaves. The unpleasant-smelling flowerheads (6 mm long, 9 mm wide) are either solitary or in clusters of three.

Leucadendron linifolium

knoppiesbos, line-leaf conebush

Female: A single-stemmed shrub up to 2 m, with twisted, linear leaves (15-35 mm long, 2 mm wide). The flowerhead has a yeasty odour and the round cones (13-24 mm wide) store seeds for several years. It grows on level, seasonally waterlogged, sandy soils over clay (wet restioid fynbos) from Bot River to Bredasdorp, as well as on the Cape Flats and at Riversdale. (Sept-Oct)

Male: Similar to the female with shorter leaves (7-27 mm long). The flowerhead has a yeasty odour.

Leucadendron elimense subsp. *elimense* ♂

Leucadendron linifolium ♂

Leucadendron laxum ♂

Leucadendron elimense subsp. *elimense* ♀

Leucadendron linifolium ♀

Leucadendron laxum ♀

Leucadendron muirii   kruiphout, luisiesbos, silver-ball conebush
Female: A straggly, single-stemmed shrub up to 2 m. The spatula-shaped leaves are thick and fleshy (*ca.* 40 mm long, 13 mm wide). Young leaves are needle-like. The dark brown involucral bracts have sharply pointed tips, and the green flowerheads have a yeast-like odour. The grey-white cones (40 mm long, 30 mm wide) contrast distinctly with the darker foliage, and in them seeds are stored for several years. It is endemic to limestone areas (lime stone proteoid fynbos) where scattered plants occur between Bredasdorp and Still Bay. (Nov to Dec)
 Male: Similar to the female except that the shrubs are denser and less straggly, with shorter leaves to 30 mm. Flowerheads are born on a 12 mm long stalk.

❑ **Leucadendron platyspermum**   kraaltolbos, platy, swartbal,
plate-seed conebush
Female: A single-stemmed, erect, bright yellow-green shrub up to 1,7 m. The bright, yellow-green leaves (*ca.* 70 mm long, 13 mm wide) contrast strikingly with the large, attractive, chocolate-brown cones (50 mm long, 40 mm wide) which have a sturdy stalk and double-edged cone bracts. Seeds are retained in cones until fire occurs. Dense stands occur in both sandy and gravelly soils in the area, as well as at Villiersdorp and Kleinmond. Cones and male foliage are extensively harvested resulting in a decline in plant numbers. (Sept)
 Male: A shorter plant than the female, up to 1,3 m tall, and with smaller leaves (up to 40 mm long, 5 mm wide).

Leucadendron salignum   duineknoppiesbos, common sunshine conebush
Female: A multi-stemmed shrub that resprouts after fire. It is usually seen as a sprawling plant less than 1 m, but is occasionally above 2 m. The leaves are hairless (*ca.* 50 mm long, 4 mm wide) and involucral leaves are ivory-coloured. Cones are round (*ca.* 20 mm wide) and retained on the plant for several years. This species occurs in many of the soils and fynbos veld types in the area, and is widely distributed between Clanwilliam and Port Elizabeth. (April to Nov)
 Male: Similar to the female, but with smaller leaves (20-47 mm long, 4 mm wide).

Leucadendron salignum ♂

Leucadendron platyspermum ♂

Leucadendron muirii ♂

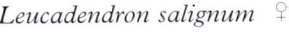

Leucadendron platyspermum ♀

Leucadendron muirii ♀

Leucadendron salignum ♀

Leucadendron salicifolium   geelbos, common stream conebush
Female: A single-stemmed, erect shrub up to 3 m. The hairless leaves are slightly sickle-shaped (60 mm long, 5 mm wide) and involucral leaves are yellow-cream. Cones (35 mm long, 25 mm wide) retain seeds for several years. It characteristically grows near streams in dense, extensive stands and occurs between Ceres and Riversdale. (July to Sept)
Male: Similar to the female.

❏ **Leucadendron stelligerum**   sterretjies, Agulhas conebush
Female: A single-stemmed shrub to 1,3 m with reddish branches. The leaves (28 mm long, 8 mm wide) have a blunt, fine tip. Like *L. modestum*, the leaves are gritty to touch, but differ in that they are covered with soft hairs. Involucral leaves are longer than the foliage leaves and surround the flowerhead in a distinctive star-like cup. The cones (23 mm long, 15 mm wide) are densely covered with hairs and are hidden by the involucral leaves. Seeds are stored for several years in these cones. This extremely rare plant occurs on gravel, ferricrete and clay soils (Elim asteraceous fynbos) from Viljoenshof to Voëlvlei. (July to Aug)
Male: Similar to the female but with slightly shorter leaves. The flowerheads have an unpleasant odour.

Leucadendron teretifolium   needle-leaf conebush
Female: A dense single-stemmed shrub up to 1 m high with needle-like leaves (22 mm long, 1 mm wide). The hairless cones (*ca.* 35 mm long, 24 mm wide) retain their seeds for many years. Plants usually grow in dense stands on shale-derived soils between Bot River and Riversdale, as well as north to the Riviersonderend and Witteberg mountains. (Aug to Sept)
Male: Similar to the female but with shorter leaves (8 mm long).

Leucadendron teretifolium ♂

Leucadendron salicifolium ♂

Leucadendron stelligerum ♂

Leucadendron salicifolium ♀

Leucadendron stelligerum ♀

Leucadendron teretifolium ♀

Leucadendron xanthoconus sickle-leaf conebush
Female: A single-stemmed shrub up to 2 m. The linear to sickle-shaped leaves have fine silver hairs when young (*i.e.* at the tips of the branches), becoming hairless later (65 mm long, 6 mm wide). Involucral leaves are yellow, and the hairy cones (30 mm long, 22 mm wide) retain seed for several years. Plants occur in large stands in acid sands of mountain slopes and coastal hills (acid sand proteoid fynbos) from the Cape Peninsula to Potberg. (Aug)
Male: Similar to the female.

In the genus *Mimetes* small clusters of florets occur above a leaf in a consecutive arrangement that forms large, cylindrical flowerheads.

Mimetes saxatilis stompie, geelstompie, limestone pagoda
A single-stemmed, erect shrub up to 2 m. The elliptic leaves (35-50 mm long, 17-30 mm wide) have one to three glandular teeth. Flowerheads (50-100 mm long, *ca.* 60 mm wide) comprise bunches of 12-22 yellow florets above a flat, green leaf. It is restricted to limestone (limestone proteoid fynbos) with a narrow distribution between Franskraal and Struisbaai. Coastal resort development is a major threat to this attractive plant. (July to Dec)

Mimetes cucullatus common pagoda, rooistompie
An erect multi-stemmed shrub up to 2 m with elliptic leaves. Flowerheads comprise bunches of 4-7 florets beneath a hooded, red-orange leaf. Found on sandstone slopes from Kouebokkeveld to the Outeniqua mountains. (Aug to Mar)

In the genus *Paranomus* the leaves are finely dissected and have grooved upper surfaces. Leaves below the flowerheads are usually quite different, being flat and undivided. Flowerheads are composed of clusters of four flowers above a leaf.

❏ **Paranomus abrotanifolius** Bredasdorp sceptre
This species is an erect shrub up to 900 mm high with a purple-pink flowerhead (*ca.* 25 mm long, 23 mm wide) comprising densely clustered florets. It occurs in sandstone-derived soils in the Elim area where it is locally common on the Elim-Wolvengat road, and also on the Potberg. There are presently only ten known populations. (May to Dec)

In the genus *Spatalla* the leaves are undivided, and the small headlets of flowers comprise either one or three florets which have prominent, spoon-like pollen presenters.

Spatalla squamata silky spoon
A single-stemmed, rounded shrub up to 500 mm with fine-tipped, linear leaves (5-12 mm long) that curve inwards. The flowerheads (20-30 mm long, 10 mm wide) consist of densely woolly headlets of one floret each, and at the base are persistent, reddish bracts with no hairs. Endemic to the Agulhas Plain area, scattered plants grow on infertile sands of coastal flats and hills (acid sand proteoid fynbos). (Aug to Oct)

Leucadendron xanthoconus ♂

Mimetes cucullatus

Mimetes saxatilis

Leucadendron xanthoconus ♀

Paranomus abrotanifolius

Spatalla squamata

❏ Spatalla ericoides
erica-leaf spoon

A single-stemmed, erect, rounded shrub up to 800 mm high, with needle-like, round-tipped leaves (7-12 mm long) that curve inwards. The densely woolly flowerheads (30 mm long, 15 mm wide) have headlets of one floret each and bracts that are red to carmine and covered with silky hairs. This species is restricted to limestone-derived sands (neutral sand proteoid fynbos) and is known from only two localities near Hagelkraal. (Aug to Oct)

Spatalla curvifolia
white-stalked spoon

A rounded single-stemmed shrub up to 0,8 m high with needle-like leaves (25-50 mm long) that have a channelled upper surface, and curve inwards. The dense, creamy-yellow flowerheads (30-70 mm long, 10-15 mm wide) comprise many headlets of one floret each. Plants occur in small stands in acid sands from Kogelberg to the Bredasdorp mountains as well as near Genadendal and Caledon. (All year)

The genus *Serruria* has dissected leaves which have cylindrical segments (unlike *Paranomus* which has grooved leaf segments). Plants have nut-like fruits that are dispersed by ants.

Serruria fasciflora
spinnekopbossie, common pin spiderhead

This species is an erect to sprawling shrub up to 900 mm, with leaves (30-70 mm long, 25-35 mm wide) that have thread-like tips bearing straight hairs. The flat-topped flowerheads (15-25 mm long, 15-50 mm wide) are pink to cream and are borne on short, hairy stalks. Plants often grow in dense populations in acid sands (acid sand proteoid fynbos) at low altitudes in the area and beyond from Ceres to George. (All year)

Serruria nervosa
fluted spiderhead

This low, erect shrub grows to 300 mm high, and has sharp-tipped leaves (20-40 mm wide, 10-20 mm long) that curve upwards. Pink, sweet-scented flowerheads on short stalks are clustered at the ends of the main branches. Scattered plants occur in low-lying limestone and acid sands between the Klein River Mountains and Elim. (July to Nov)

Serruria elongata
long-stalk spiderhead

An erect shrub growing to 1,5 m high, with upward-curving leaves (50-150 mm long, 60 mm wide) that have rounded tips with fine points. The silky-haired, pink flowerheads are sweetly scented and have characteristic long stalks up to 300 mm long. It grows on acid flats and slopes from Du Toit's Kloof to Elim. (Aug to Dec)

Serruria bolusii

This plant is so similar to *S. nervosa* that it is frequently thought to be the same species. However, it differs in that the flowerheads do not all occur at the ends of the main branches, but arise at intervals along it from smaller side branches. It is restricted to the Bredasdorp area where it occurs in acid sands at Quoin Point and on the hills near Elim. (July to Nov)

Spatalla curvifolia

Spatalla ericoides

Serruria fasciflora

Serruria bolusii

Serruria nervosa

Serruria elongata

The genus *Leucospermum* comprises species that have rounded, compact flowerheads with stout, relatively long, protruding styles resembling a "pincushion". Seeds fall to the ground at the end of the season and are carried underground by ants.

❑ **Leucospermum truncatum**   waboom, limestone pincushion
A single-stemmed, erect shrub up to 2 m. The wedge-shaped leaves (45-90 mm long, 8-15 mm wide) have flattened ends with three glandular teeth. Flowerheads (*ca.* 40 mm wide) are golden-yellow later turning orange. Plants occur in dense, isolated stands in limestone areas (lime stone proteoid fynbos) between Bredasdorp and Riversdale. (Aug to Dec)

Leucospermum patersonii   basterkreupelhout, silver-edge pincushion
A single-stemmed shrub up to 4 m. The oblong leaves (50-90 mm long, 40 mm wide) have three to eight glandular teeth and the flowerheads are orange-crimson (85 mm wide). It is endemic to limestone areas (limestone proteoid fynbos) between Hermanus and Cape Agulhas. (Aug to Dec)

❑ **Leucospermum fulgens**   Potberg pincushion
A single-stemmed, rounded shrub up to 3 m. The oblong leaves (60-90 mm long, 15-20 mm wide) are hairy when young, becoming hairless on maturity and have two to four glandular teeth. The stalkless flowerheads (*ca.* 70 mm wide) are pink to orange becoming crimson when mature. It is restricted to a small patch of neutral sand proteoid fynbos between the limestone hills and the Potberg range in the De Hoop Nature Reserve. Much of its habitat is overrun by alien vegetation. (Aug to Jan)

Leucospermum cordifolium   luise, speldekussing, "The" pincushion
A single-stemmed, rounded shrub up to 1,5 m high with horizontally drooping branches. The leaves (20-80 mm long, 20-45 mm wide) have a rounded tip with up to six glandular teeth. Flowerheads (100-200 mm wide) are yellow, orange or crimson and are borne at right angles to the stem. A well-known species, it occurs in acid sand proteoid fynbos in the Bredasdorp mountains, Soetanysberg and Elim areas, as well as at Kleinmond and Houhoek. (Aug to Jan)

Leucospermum cuneiforme   wart-stemmed pincushion
A multi-stemmed shrub up to 3 m that has characteristic warts and pustules at the bases of its stems at ground level. The wedge-shaped leaves (45-110 mm long, 6-30 mm wide) have three to ten glandular teeth. Flowerheads (50-90 mm wide) are yellow, becoming orange as they mature. A few scattered plants occur on the Potberg in acid sand proteoid fynbos, and also elsewhere between Riviersonderend and Grahamstown. (All year, mainly Aug to Feb)

Leucospermum cordifolium

Leucospermum fulgens

Leucospermum truncatum

Leucospermum cuneiforme

Leucospermum patersonii

Leucospermum hypophyllocarpodendron subsp. **hypophyllocarpodendron**   slangbossie, green snake-stem pincushion
A prostrate shrub 200 mm high with long, horizontal, trailing stems forming mats up to 1,5 m wide. The leaves (40-130 mm long, 2-15 mm wide) point directly upwards and the bright yellow flowerheads (*ca.* 35 mm wide) occur in groups of about four. Plants grow in sandy soils of flats and lower slopes from the Cape Peninsula to Bredasdorp. (Aug to Jan)

❏ **Leucospermum heterophyllum**   rankluisie, trident pincushion
A low (150 mm high), ground-hugging shrub forming large mats up to 5 m wide. The stalkless leaves (*ca.* 25 mm long, 4 mm wide) have one to three glandular teeth. Young leaves are hairy, later becoming hairless. Flowerheads (25 mm wide) are yellow-green with styles that turn carmine when mature and are borne on 15 mm long stalks. It occurs in sandstone or gravelly, clayey soils from Elim to De Hoop. (Aug to Jan)

❏ **Leucospermum pedunculatum**   white trailing pincushion
A single-stemmed, prostrate shrub up to 300 mm high forming wide (up to 3 m), dense mats. The bright, green leaves are linear (30-60 mm long, 2-5 mm wide) and generally point upright from the stout, horizonal branches. The small, (30 mm wide) sweet-smelling flowerheads are creamy-white, later turning to carmine. It occurs in deep sands in a narrow, coastal strip from Danger Point to Soetanysberg. (Aug to Jan)

Leucospermum prostratum   yellow trailing pincushion
A prostrate shrub with many long, slender, trailing branches arising from an underground rootstock, forming a mat up to 4 m wide. The linear leaves (*ca.* 30 mm long, 2-6 mm wide) have pointed tips, and the dainty (25 mm), sweet-smelling flowerheads are bright yellow turning deep orange. This species occurs in deep sands, mainly in coastal areas, from the Kogelberg to the Elim hills. (July to Dec)

Leucospermum truncatulum   buxifolium, oval-leaf pincushion
A single-stemmed, sparsely branched, erect shrub up to 2 m. The elliptic leaves (10-25 mm long, 5-10 mm wide) are densely hairy and overlap each other on the stems. Small, yellow flowers (*ca.* 20 mm wide) occur in clusters of two to eight and become crimson when mature. It occurs in acid sand proteoid fynbos over extensive areas on the Bredasdorp Mountains and Soetanysberg, and also further west to the Kogelberg. (Aug to Dec)

Leucospermum hypophyllocarpodendron subsp. *hypophyllocarpodendron*

Leucospermum heterophyllum

Leucospermum pedunculatum

Leucospermum truncatulum

Leucospermum prostratum

Protea species have several rows of overlapping petal-like involucral bracts, usually brightly coloured, surrounding the central head of numerous, small florets.

Protea aspera aardroos, snowball, rough-leaf sugarbush
A low, ground-flowering spreading shrub up to 200 mm high with rough flat, linear leaves (70-200 mm long, 3-14 mm wide). Flowerheads (65-95 mm long, 50 mm wide) are surrounded by hairy golden-brown involucral bracts. Scattered plants occur in acid sand proteoid fynbos on flats and slopes in the southern Overberg and also in the Langeberg. Flowering is mainly after fire. (Sept to Dec)

Protea subulifolia awl-leaf sugarbush
A low shrub up to 700 mm tall with upright stems. Leaves vary in size (5-85 mm long, 0.5-2 mm wide) and shape (needle-like to round and grooved). The ground-level, cup-shaped flowerheads (*ca.* 50 mm wide) have a yeasty odour that attracts rodent pollinators. Involucral bracts are brown to pink and densely hairy. Scattered plants occur in sandy to clay soils from Bot River to Elim, as well as further inland from the Stettynskloof to the Langeberg mountains. In lowland areas the species is threatened by agriculture and plants are confined to road verges. (July to Oct)

❑ **Protea pudens** minor, skaamgesiggie, bashful sugarbush
A single-stemmed, spreading prostrate shrub up to 400 mm tall with linear leaves (60-140 mm long, *ca.* 3 mm wide) point upwards from trailing stems. Dainty, bell-shaped flowerheads (50-80 mm long, 30-60 mm wide) are hidden in the foliage, and have a central, woolly cone and rusty pink, fringed involucral bracts. Found in the clay-rich soils of Elim asteraceous fynbos only, a habitat widely used for agriculture. Despite their overall decline, they are locally abundant, especially on Geelkop near Elim. (May to Sept)

Protea denticulata tandjies, tooth-leaf sugarbush
A dense shrub up to 1 m tall, with linear, channelled, upward-curving leaves (*ca.* 200 mm long, 6 mm wide) with horny denticles. Tiny flowerheads (*ca.* 45 mm long, 38 mm wide) grow close to the ground and have reddish-carmine velvety involucral bracts. Plants are restricted to acid sand proteoid fynbos on the Potberg. (Aug to Oct)

❑ **Protea aurea** subsp. **potbergensis** Potberg protea/sugarbush
A single-stemmed, erect shrub up to 5 m. Leaves are oval with a heart-shaped base (35-75 mm long, 25-55 mm wide). Flowerheads (90-130 mm long) have creamy-green involucral bracts and when fully open resemble a shuttle cock. This subspecies is confined to shale bands on the Potberg. *Protea aurea* subsp. *aurea* is common on the Langeberg and Outeniqua mountains. (All year, mainly Jan to June)

Protea speciosa brown-bearded protea/sugarbush
A sturdy, erect-stemmed, shrub growing to 1,2 m from a fire-resistant rootstock. Leaves (90-160 mm long, 10-60 mm wide) have thickened margins and flowerheads (90-140 mm long, 70 mm wide) have closely packed pink to brown, fringed involucral bracts. Scattered plants grow in acid sand proteoid fynbos from the Cape Peninsula to Bredasdorp. (June to Jan)

Protea aspera

Protea subulifolia

Protea pudens

Protea aurea subsp. *potbergensis*

Protea speciosa

Protea denticulata

Protea susannae stink-blaar(leaf)protea/sugarbush
A single-stemmed shrub up to 3 m, with wavy leaves (80-160 mm long, 15-30 mm wide) that have an unpleasant, sulphurous odour when crushed. The pink flowerheads (80-100 mm long, 70-110 mm wide) have a darkish, sticky film on their bracts, and seeds are retained in seedheads for many years. It occurs in dense stands in neutral sand proteoid fynbos from Gansbaai to the Gouritz River. The leaves have been used for tanning skins. (April to Sept)

Protea compacta compacta, duinesuikerbos, Bot River protea/sugarbush
A single-stemmed, erect and sometimes untidy shrub up to 5 m tall. The oval leaves are heart-shaped at the base (50-130 mm long, 20-55 mm wide) and flowerheads (90-120 mm long, *ca.* 60 mm wide) have pink involucral bracts that are fringed with hairs. Seeds are retained in seedheads until fire occurs when they are released. They occur as large, dense stands in acid sand proteoid fynbos of both mountainous and flat areas between Kleinmond and Bredasdorp. The attractive flowers are harvested in large quantities by the wild flower industry. (April-Sept)

Protea longifolia swartbaard, wolkop, long-leaf sugarbush
A sprawling shrub up to 1.5 m high. It has upward-curving leaves (90-200 mm long, 5-17 mm wide) and greenish, pink or white flowerheads (80-160 mm long, 40-90 mm wide) that have a central, pointed, black woolly cone. Found in acid sand proteoid fynbos on the lower slopes of the Bredasdorp Mountains and the Soetanysberg, and also on the Du Toit's Kloof and Riviersonderend Mountains. A small form, *P. longifolia* var. *minor*, is found in Elim asteraceous fynbos. This variant grows to 0.5 m and has small, green flowerheads about 120 mm long. (May to Sept)

Protea obtusifolia Bredasdorp protea, limestone sugarbush
A single-stemmed, rounded and erect shrub growing to 2 m, with upward curving, leathery leaves (*ca.* 130 mm long, 30 mm wide) and red, occasionally cream, flowerheads (90-120 mm long, 50-80 mm wide). The pointed, cone-shaped seedheads retain seeds for many years. Endemic to limestone proteoid fynbos in the area where dense stands are formed and often grows together with *Leucadendron meridianum*. It is also found further east towards Riversdale. (April to Sept)

Protea repens suikerbos, common sugarbush
Up to 2 m high, with upright leaves (50-100 mm long, 5-18 mm wide). Flowerheads (100-160 mm long, 80 mm wide) have pointed, sticky, cream to pinkish-red bracts, forming cone-shaped seedheads that remain on the plant for several years. This widespread species occurs in many different habitats from Nieuwoudtville to near Grahamstown. The copious nectar produced by the flowers is loved by baboons, and was widely used for sugar and syrup (bossiestroop) production by early settlers. (All year)

Protea longifolia

Protea longifolia (small flowered form)

Protea obtusifolia

Protea susannae

Protea repens

Protea compacta

VISCACEAE

Viscum capense voëlent
A leafless, woody parasite with fleshy, jointed stems. Flowers are inconspicuous and berries ripen in spring to a translucent yellow. It is often found growing on species of *Rhus*. (June to Oct)

SANTALACEAE
Members of this family are partially parasitic on the roots of other plants. Some species have reduced leaves and are yellowish due to low levels of chlorophyll.

Colpoon speciosum
A rigid, sparsely-branched shrub up to 2 m high that sprouts after fire, with oval, dull-green leaves (25-46 mm long; 10-26 mm wide). Numerous upright stems bearing clusters of small, yellowish-green, rank-smelling flowers are produced within the first weeks after fire. The fleshy fruits (25 mm long, 21 mm wide) are yellow turning purple-black, and are relished by baboons. Flowering occurs only in the first season after fire. It occurs on lower mountain slopes, and on deep sands at the base of limestone hills from the Hottentots Holland mountains to Agulhas.

Colpoon compressum (= *Osyris compressa*) basbessie
A dense shrub or small tree up to 3 m high that resprouts after fire. It has grey-green oval leaves (*ca.* 30 mm long) and small, whitish-yellow flowers and edible, dark red-black, fleshy fruits (17 mm long, 12 mm wide). It occurs on coastal and inland slopes from the Cederberg to Port Elizabeth as well as in the interior. (Aug to Nov)

Thesidium fragile
An erect, almost leafless shrublet up to 500 mm high with minute, round, white flowers. A fleshy cup covers the base of the small fruits. It grows on sand dunes from the Peninsula to Riversdale. (All year)

Thesium capitatum
A shrublet up to 300 mm tall with small, triangular, closely packed leaves. The minute, whitish, star-like flowers are grouped at the ends of branches. It occurs on mountain slopes from Malmesbury to Humansdorp. (All year)

Thesium penicillatum
A stout, sparsely branched shrub up to 1 m tall with small, whitish flowers. It occurs on mountain slopes from Caledon to Humansdorp. (Sept to Feb)

Viscum capense

Colpoon speciosum

Colpoon compressum

Thesium capitatum

Thesidium fragile

Thesium penicillatum

GRUBBIACEAE

Grubbia rosmarinifolia skilpadbossie
A much-branched shrub up to 1,5 m high with narrow, ericoid leaves and clusters of small, woolly, white to pink flowers. Plants occur on marshy slopes from the Cederberg to the Cape Peninsula and Uniondale. (Aug to Nov)

CHENOPODIACEAE

Bassia diffusa (=*Chenolea diffusa*) soutbossie
A silver, mat-like perennial with reddish stems and small grey leaves (*ca.* 6 mm long) covered with silky hairs. The flowers are inconspicuous. It grows among rocks in the salt-spray zone and in estuaries from Saldanha to Natal. (Feb to April)

Sarcocornia littorea koraalbrak/bos, lidjiesbos
An erect woody shrub up to 1 m high with fleshy, jointed branches and minute, hidden flowers. It grows on rocky coastal areas, where it is subjected to occasional inundation, between the Cape Peninsula and Port Elizabeth. (Sept to Dec)

AIZOACEAE

Aizoon rigidum var. **angustifolium**
A creeping shrublet that forms dense mats on disturbed ground. It has tough stems with soft, thin, greyish leaves, and small yellow flowers. Distribution extends from Caledon to Port Elizabeth. (Sept to Oct)

Tetragonia herbacea
A ground-hugging, soft, perennial plant with a tuberous root. The leaves are slightly fleshy and the yellow flowers have four petals. Fruits are pear-shaped. It is common in sandy areas from Clanwilliam to Bredasdorp and also occurs in Namaqualand. (May to Sept)

Tetragonia decumbens
A glistening, perennial plant with a thick, tough, creeping stems. The fleshy leaves are covered with shiny glands and the small, yellow flowers have four petals. Fruits are winged. It grows in coastal sands from the Cape Peninsula to Port Elizabeth as well as in Namaqualand and Namibia. (August to Mar)

Grubbia rosmarinifolia

Bassia diffussa

Sarcocornia littorea

Aizoon rigidum var. *angustifolium*

Tetragonia decumbens

Tetragonia herbacea

MESEMBRYANTHEMACEAE

❏ Caryotophora skiatophytoides
A perennial, herbaceous prostrate plant that is only slightly succulent. It has spreading branches and longish, flat, fleshy leaves (*ca.* 160 mm long and 30 mm wide). The white flowers are about 60 mm wide. This plant is endemic to seasonally waterlogged flats around the Soetanysberg and is conspicuous only after fire. (Oct to Nov)

Carpobrotus acinaciformis   suurvy, sour fig
A mat-like succulent with trailing stems. The large, scimitar-shaped leaves are sharply three-angled and occur in pairs, slightly united at the base. The large, showy flowers are rose-purple and occur singly at the ends of a shortish stalk. Fruits are fleshy, constricting at the base where they join the stalk and become tough and leathery. This species occurs in coastal sands between the Cape Peninsula and Port Elizabeth. The raw leaves were chewed as a cure for sore throats. (Aug to Oct)

Carpobrotus edulis subsp. edulis   hottentotsvy/fig
This plant superficially resembles *C. acinaciformis* but differs as follows: flowers are yellow or pink with two of the calyx lobes longer than the petals; the leaves are straight to slightly curved with serrated margins towards the tips; the fruit tapers at the base into the stalk. The pulpy fruits are edible and are used to make a jam. It is widespread in the south-western and eastern Cape areas. Like *C. acinaciformis* the leaves have medicinal properties and the expressed juice has been used as a lotion for blue-bottle stings and burns. (Aug to Oct)

Conicosia pugioniformis subsp. muirii   volstruisvygie
A spreading, perennial plant up to 200 mm high with erect tufts of long, narrow, succulent leaves that are triangular in cross-section. The dark yellow flowers (*ca.* 90 mm wide) occur singly at the end of long stalks. Fruit capsules are cone-shaped and consist of 12-18 narrow segments that are separated at the top and do not spread when moistened. This species occurs from Somerset West to Knysna. (Aug to Sept)

Carpobrotus acinaciformis

Carpobrotus edulis subsp. *edulis*

Caryotophora skiatophytoides

Conicosia pugioniformis subsp. *muirii*

The following Mesembryanthemaceae species have segmented, woody fruit capsules that open on wetting and close when dry. In this way the enclosed seeds are partially released when it rains, and protected in the dry intervening periods.

Delosperma litorale kalkklipvygie
An erect or creeping shrublet with trailing stems, growing to 150 mm tall and forming mats up to 350 mm wide. The flowers may be solitary, or may occur in groups and range from white or cream to mauve in colour. Fruit capsules have five segments. It occurs in limestone areas from Bredasdorp to Port Elizabeth. (April)

Dorotheanthus bellidiformis sandvygie
An annual, herbaceous plant up to 100 mm tall. The flat, lance-shaped, fleshy leaves are papillate (have fluid-filled, bladder-like cells on the surface) and form loose, tufted rosettes. The flowers (*ca.* 40 mm wide) may be white, pink, red or brilliant purple. Fruit capsules have five segments. It occurs in sandy areas from Clanwilliam to Bredasdorp. (Aug to Sept)

Drosanthemum hispidum
An erect or spreading shrublet up to 600 mm high with cylindrical pairs of succulent leaves (*ca.* 20 mm long) that are covered with glistening papillae (fluid filled, bladder-like cells) on their surfaces. The white to pink or mauve flowers (*ca.* 3 cm wide) occur singly at the end of a short stalk. Fruit capsules have four to six segments. It grows in sandy or clayey, usually dryish areas between Piketberg and Humansdorp. (Sept to Jan)

Drosanthemum intermedium
A sparsely branched, low shrublet with slender, creeping stems that are thickened at the nodes, and succulent, papillate leaves like the above species. The mauve to magenta flowers (*ca.* 25 mm wide) occur singly on short stalks and the fruit capsules have four to six segments. It occurs usually in coastal sandy areas between Malmesbury and Port Elizabeth. (Aug to Sept)

Lampranthus amabilis
This sprawling plant (*ca.* 200 mm high) has small leaves (8-12 mm long) and orange to magenta flowers (10 mm wide). Fruit capsules have five segments. It grows on coastal sands and limestone, as well as on low stony slopes in the southern Overberg. (Sept)

Ruschia geminiflora
A mat-forming plant (600-900 mm wide) often forming roots along the stems. The grey-green, succulent leaves (20-50 mm long, 4 mm wide) are covered with translucent dots and are 3-angled with toothed margins. The flowers are bright pink (5-10 mm wide). Fruit capsules have at least five segments. It occurs on sandy areas from Saldanha Bay to De Hoop. (July to Oct)

Delosperma litorale
Dorotheanthus bellidiformis

Ruschia geminiflora   *Lampranthus amabilis*

Drosanthemum hispidum
   Drosanthemum intermedium

Glottiphyllum depressum tongblaarvygie
A small perennial plant branching close to the ground such that it appears stemless. The densely packed, fat leaves are soft, pulpy and indented, resembling tongues. The solitary flowers are bright yellow and the fruit capsules have 8-20 segments. It is locally common in certain clayey habitats in the area and also occurs near Humansdorp. (July to Aug)

Jordaaniella dubia
A mat-forming succulent with ground-creeping stems that root at the nodes. The cylindrical or semi-cylindrical leaves are 25-30 mm long and the bright yellow flowers 25-30 mm across. Fruit capsules have 8-20 segments. It grows on deep sand between limestone ridges, and also on coastal dunes from Lamberts Bay to Agulhas, as well as in Namaqualand and Namibia. (June to Aug)

Prenia vanrensburgii seepampoen
A prostrate, succulent plant with long, creeping stems, forming mats several metres in diameter. The fat, ovate leaves (30-40 mm long, 20 mm wide) have a thick, waxy covering that can easily be wiped off. The white-yellowish flowers (up to 40 mm wide) occur singly or in loose clusters at the ends of stems and the fruit capsules have four to five segments. This species is endemic to the Bredasdorp area and occurs on rocky coasts, often close to the high water mark. (October)

Jordaaniella dubia

Glottiphyllum depressum

Prenia vanrensburgii

CARYOPHYLLACEAE

Dianthus albens
wilde-angelier, wild pink

A tufted or spreading herbaceous perennial up to 300 mm high with narrow leaves (*ca.* 25 mm long). The solitary flowers (*ca.* 300 mm wide) are white with a pink centre, or pink with a white centre and the petals have blunt, ragged ends. The calyx is tubular with several overlapping bracts at the base. It grows on sandy flats and slopes from the Cape Peninsula to Mossel Bay. (Nov to April)

Silene undulata
wildetabak

A hairy, often sticky annual or perennial plant 150 to 800 mm high with pairs of pointed leaves (*ca.* 200 mm long). The white, or pinkish-red, flowers (*ca.* 300 mm wide) have bilobed petals and a long, tubular calyx. Found in shady places among bushes and rocks throughout most of southern Africa. (July to Sept)

NYMPHACEAE

Nymphaea nouchali var. caerulea (= *N. capensis*)
blouwaterlelie, water/lotus lily

An aquatic, perennial plant with large, round leaves that float on the surface of the water. The showy, blue to lilac flowers (*ca.* 150 mm wide) are sweet-smelling, and have four sepals and many petals. Occasionally seen in pools and vleis from the Cape Peninsula to Port Elizabeth, and also further up the coast to tropical Africa and Madagascar. (Sept to Oct)

RANUNCULACEAE

Anemone tenuifolia
wild anemone

A herbaceous perennial plant growing up to 400 mm. The rather stiff basal leaves are divided into sharp-tipped lobes and the soft, white to pink flowers (*ca.* 90 mm wide) are borne on long, furry stalks. It grows on moist upper slopes from Piketberg to Port Elizabeth. (June to Feb)

Knowltonia anemonoides
A perennial plant up to 900 mm high with large, basal leaves divided into oval lobes. Clusters of smallish, white to greenish-yellow flowers (*ca.* 20 mm wide) occur at the ends of long stalks and the fleshy fruits are small and dark. It grows in shady, wooded habitats on mountain slopes between the Cape Peninsula and Riversdale. (Sept to Jan)

Dianthus albens

Silene undulata

Knowltonia anemonoides

Nymphaea nouchali var. *caerulea*

Anemone tenuifolia

MENISPERMACEAE

Cissampelos capensis
A woody, climbing plant that sprawls over bushes. Leaves are heart-shaped (*ca.* 24 mm long) and the green flowers are inconspicuous. Found on stony slopes from Namaqualand to Port Elizabeth and also in the Little Karoo. (Feb to May)

LAURACEAE

Cassytha ciliolata
nooienshaar, false dodder

A yellow-green, leafless, twining parasite with no roots and a tangle of stems that have suckers for attachment to host plants. The small, yellowish-white flowers occur in clusters and produce red or yellow berries. It occurs between the Cape Peninsula and Port Elizabeth. (All year)

BRASSICACEAE
Plants in the Brassicaceae family have four petals that spread outwards in the form of a cross.

Heliophila subulata
A sprawling herb or a soft shrublet varying between 100 and 500 mm in height. It has narrow, fleshy leaves (10-70 mm long) and blue-mauve to pink flowers. Distribution is between the Cape Peninsula and Port Elizabeth, as well as further east to Kwazulu-Natal. (Mar to Sept)

Heliophila macra
A slender, sparingly branched shrub up to 1,3 m high with linear, pointed leaves (25-80 mm long). Flowers are white or mauve. It occurs in sandy areas between Swellendam and the southern Overberg. (Oct to April)

DROSERACEAE

Drosera capensis
sundew, doublom, snotblom

Fairly sturdy, sticky plants up to 300 mm high. The long, linear leaves (*ca.* 150 mm long; 4 mm wide) have knob-shaped tentacles that exude mucilaginous drops which trap and digest insects. Reddish-purple to pink or mauve flowers (*ca.* 30 mm wide) occur in loose clusters. It occurs in damp areas between Clanwilliam and Port Elizabeth. (Nov to Jan)

Drosera cistiflora
slakblom

A slender, sticky plant up to 400 mm, with a basal rosette of tentacle-bearing leaves (12-20 mm long) as well as longer leaves arising from the upright stem. The leaves have tentacles that exude mucilaginous drops and flowers may be white, yellow, pink to purple and have a dark green centre. Commonly found in damp areas from Clanwilliam to Port Elizabeth. (Aug to Sept)

Cissampelos capensis

Cassytha ciliolata

Heliophila macra

Drosera capensis

Drosera cistiflora

Heliophila subulata

CRASSULACEAE

Adromischus caryophyllaceus nenta(bossie)
A branched, fleshy shrublet about 250 mm tall with unspotted, oblong leaves (*ca.* 25 mm long) and a spike of stalkless flowers borne on a tall, protruding stem. Flowers have a narrow green tube flaring out into white, pink or purple lobes that sometimes have darker markings. It occurs on rocky outcrops in renosterveld in the southern Overberg and also eastwards to Mossel Bay, as well as in the Little Karoo. (Oct to Mar)

Cotyledon orbiculata var. **orbiculata** plakkie, hondeoor
A shrubby plant up to 1 m tall with stout, succulent stems and leaves. The large, oval ear-like leaves (*ca.* 100 mm long) have red margins. The stalked, hanging tubular flowers have backward-curled petals at the rim, and are dusky pink-rose to scarlet. It is common on dunes in the area and is also found elsewhere in dry regions of southern Africa. The leaves were used to treat corns and fever blisters. (All year)

Crassula expansa subsp. **filicaulis**
A prostrate, succulent shrub with stilt-like roots along the branches and fleshy, oblong leaves (6-20 mm long, 2 mm wide) that are sometimes red. Small white to pink to yellow flowers (*ca.* 3 mm wide) occur singly along the stems. Plants occur on coastal dunes and other sandy habitats and may also be found growing in the sandy pockets formed in limestone bedrock in the area. This widespread species is found from the Cape Peninsula to Port Alfred and also along the west coast to southern Namibia. (All year)

Crassula nudicaulis var. **nudicaulis**
A tufted plant up to 300 mm high with oblong, succulent leaves (50-80 mm long, 6-15 mm wide) and greenish-white flowers. This widespread species occurs in sandy and gravelly soils from Clanwilliam to Port Elizabeth as well as further afield to the Free State and northern Natal. (July to Dec)

Crassula fallax
Up to 400 mm high when in flower, with ground-hugging branches that have many leaves at the base and old leaves remaining further up the stem. The oblong leaves (30-50 mm long, 4-10 mm wide) are slightly fleshy and have a hairy margin. Heads of whitish-cream flowers are borne at the end of branches. It occurs on sandy slopes in the area, as well as between the Cape Peninsula and Clanwilliam. (Nov to Mar)

Crassula nudicaulis var. *nudicaulis*

Crassula fallax

Cotyledon orbiculata var. *orbiculata*

Crassula expansa subsp. *filicaulis*

Adromischus caryophyllaceus

BRUNIACEAE

Berzelia abrotonoides rooibeentjies
A shrub up to 1,5 m high with small, needle-like leaves. The cream flowers occur as compact, round balls arranged in a branched head. On close inspection a single style can be seen in each tiny, individual flower, a feature which distinguishes this genus from similar-looking *Brunia* species which have two styles. Found in marshy areas on flats, this species occurs from Clanwilliam to Port Elizabeth. (April to Oct)

Berzelia lanuginosa kolkol
A shrub up to 2 m high with needle-like leaves on soft, hanging branches. Dense, round balls (*ca.* 7 mm wide) of flowers occur in a branched head. It occurs on sandy flats and slopes in permanently moist sites between Clanwilliam and Bredasdorp. (June to Nov)

Brunia laevis vaalstompie, brunia, vaaltol
A sturdy shrub up to 900 mm high with small, overlapping, grey leaves. The flowerheads comprise compact balls of creamy flower heads (*ca.* 15 mm wide). It grows on lower to middle slopes in the Caledon to Bredasdorp area. (Aug to Jan)

Nebelia paleacea
A highly-branched shrub up to 800 mm high with small needle-like leaves. The balls of cream flowers (*ca.* 7 mm wide) are surrounded by pointed bracts and are arranged in a branched head. It occurs on mountain slopes between Clanwilliam and George. (Oct to Feb)

Staavia radiata altydbossie
A shrub up to 800 mm tall with slender branches bearing small, needle-like leaves. The small flowerheads (*ca.* 7 mm wide) occur either singly or in loose groups, and look like "daisies" with an outer ring of white bracts surrounding a central disc of tiny, pink-purple flowers. The all year flowering season is the basis for its common name (= the "always" bush). It grows on sandy flats and lower slopes from Malmesbury to Riversdale. (Oct to Feb)

Berzelia abrotanoides

Brunia laevis

Berzelia lanuginosa

Staavia radiata

Nebelia paleacea

ROSACEAE

Cliffortia feruginea pypsteelbos
A sprawling shrublet with upright stems and simple, toothed and shiny leaves that have small, bract-like lobes (stipules) at their point of attachment to the stem. The minute white to pinkish male and female flowers occur on different plants. It is found on flats and lower slopes, usually near water, from the Cape Peninsula to Port Elizabeth. The branches were used for pipe stems. (Nov to July)

Cliffortia strobilifera
An erect shrub growing to 3 m high with clusters of narrow, linear leaves and inconspicuous, yellowish flowers. The cone-like bodies often seen on the plant are insect-induced galls. This species is widespread in marshy areas of the Western Cape, Namaqualand and from the Eastern Cape and further inland. (Jan to Mar)

Cliffortia ilicifolia var. **ilicifolia**
An erect shrub growing to 2 m high with simple, flat leaves that have hard, spiny teeth on their margins. It occurs on damp sandstone slopes between the Cape Peninsula and Port Elizabeth. (Mar to Dec)

Cliffortia stricta
An upright shrub up to 1,5 m high with tufts of ericoid leaves 4-8 mm long. The minute, purple male and female flowers occur on separate plants. This species grows on flats and lower slopes between the Cape Peninsula and Port Elizabeth and is fairly common on the Heuningberg where it grows in association with *Protea compacta* and *Leucadendron xanthoconus*. (Oct to June)

Cliffortia ferruginea

Cliffortia ilicifolia var. *ilicifolia*

Cliffortia stricta ♀

Cliffortia strobilifera

Cliffortia stricta ♂

FABACEAE

The cosmopolitan Fabaceae is the second largest family of flowering plants. It comprises trees, shrubs and many kinds of herbaceous plants. The fruit is usually a legume (pod) which splits to release the seeds.

Acacia cyclops   rooikrans

This one of many Australian *Acacia* species that were introduced to South Africa and are now a major threat to the Cape flora. The lack of natural predators in their adopted home has led to their rapid spread, and the most effective form of stopping such invasions has been the introduction of specific biological agents (*eg.* insects, fungi) from their home country.

Acacia cyclops is a tree growing up to 6 m and can be recognised by its persistent, empty, curled pods. The black seeds are encircled with fleshy, red structures (arils) to attract birds, their major dispersers. It has formed dense thickets in many dune fynbos and neutral sand proteoid fynbos areas. The wood makes excellent firewood. A seed-eating beetle has been introduced as a biological control agent. (Nov to Jan)

Acacia longifolia   langblaar

A tree up to 7 m, with long leaves (40-180 mm) and conspicuous parallel veins. The yellow flowerheads occur in a spike, and the pods are straight with constrictions between the seeds. It grows along streams as well as on drier flats and slopes. Spread has been successfully curbed by a wasp which causes gall formation in developing flowers, thus preventing the production of seed. These large, often reddish galls are so prevalent that they are now a characteristic feature. (June to Nov)

Acacia mearnsii   swartwattel, black wattle

A large tree up to 15 m, with finely divided, bipinnate leaves and round heads of yellow flowers. It invades along water courses and causes a serious reduction in water flow. A seed-eating beetle has been deployed to control the spread of this species. Biological agents that destroy the plant cannot be used as this tree is commercially grown for its tannin-rich bark. (Aug to Nov)

Acacia saligna   Port Jackson

A large tree up to 9 m high with round, yellow flowerheads. It forms extensive thickets in riverine areas as well as on acid, sandy flats and lower slopes. The introduction of a fungus which impairs growth and ultimately kills this species is hoped to curb its spread. (Aug to Oct)

Acacia cyclops

Acacia lougifolia (with galls)

Acacia mearnsii

Acacia saligna

Amphithalea alba
A dwarf, straggling shrub with erect branches up to 300 mm high. It has small, simple silvery leaves (*ca.* 8 mm long) and the small white to cream flowers are crowded into a rounded head at the ends of branches. It occurs in the Bredasdorp and Riversdale areas. (May to Aug)

Amphithalea biovulata
A delicate trailing shrublet up to 600 mm high with scale-like leaves. Small mauve flowers occur at the ends of branches. It occurs in the Caledon and Bredasdorp areas and is common in Elim fynbos. (Sept to Oct)

Amphithalea ericifolia subsp. ericifolia
A sprawling shrub up to 600 mm high with slender, erect branches and shiny, ericoid leaves. The small pink to mauve flowers occur in dense spikes towards the ends of branches. It is common in Elim fynbos as well as further afield towards the Cape Peninsula and Riversdale. (June to Sept)

Aspalathus calcarea
A densely leafy, prostrate to upright shrub up to 600 mm tall. The young branches are woolly. The needle-like leaflets are 2-6 mm long, and yellow flowers are born on short side branches. This species is restricted to limestone soils, and occurs between Bredasdorp and Riversdale. (April to Oct)

Aspalathus caledonensis
An erect, rod-like shrub up to 1,2 m tall with flat silvery leaves. The central leaflet (*ca.* 5 mm long, 2 mm wide) is about twice the size of the two side leaflets. It has lemon-yellow flowers arranged in a spike-like fashion along much of the plant's length. This species occurs on sandy or gravelly slopes of the mountains in the area. It is common in the Heuningberg Nature Reserve. (Aug to Nov)

Aspalathus caledonensis   *Amphithalea ericifolia* subsp. *ericifolia*

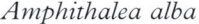

Amphithalea alba   *Aspalathus calcarea*

Amphithalea biovulata

Aspalathus hirta subsp. **hirta**
An erect, branched, spiny shrub up to 2 m high with hairy branches and needle-like, partially-hairy leaflets (*ca.* 12 mm long). The calyx and upper petals of the bright yellow flowers are also hairy. It grows on sandy flats and lower slopes of acid proteoid fynbos as well as in renosterveld between the Gifberg and Uniondale. (Sept to Dec)

Aspalathus incurvifolia
A sprawling to upright, shrub up to 1 m high with long, densely leafy branches. The leaflets are linear (5-18 mm long, *ca.* 0.5 mm wide) and the small, yellow flowers are clustered towards the ends of the branches. This species is common in limestone areas as far east as Mossel Bay. (June to Sept)

Aspalathus juniperina
A prostrate or spreading shrub 200-900 mm high with furry young branches. The leaflets (*ca.* 10 mm long) are needle-like and the yellow flowers are often tinged reddish purple, and usually clustered at the tips of branches. It occurs in acid, sandy soils between Piketberg and Bredasdorp. (Oct to Mar)

Aspalathus pycnantha
A dense, rounded shrub up to 500 mm tall with lax branches and silvery-furry leaflets (*ca.* 6 mm long). Solitary, pale yellow flowers (*ca.* 6 mm long) grow along the branches. This species is endemic to the Bredasdorp area and grows in fynbos on flats and lower mountain slopes, as well as in renosterveld. (July to Oct)

Aspalathus securifolia
A fairly robust, erect shrub up to 1,2 m tall with young branches that are furry. The leaflets are broadly oval (*ca.* 20 mm long, 4 mm wide) and there are terminal clusters of six to ten yellow flowers on slender, furry stems about 5 mm long. It grows on lower gravelly or rocky mountain slopes in this area and towards Riversdale, as well as in the Worcester region. (All year)

Aspalathus hirta subsp. *hirta*

Aspalathus juniperina

Aspalathus incurvifolia

Aspalathus pycnantha

Aspalathus securifolia

Coelidium ciliare
An erect or spreading shrublet up to 300 mm tall with narrow leaves (10-12 mm long, 0,5 mm wide). The small (*ca.* 10 mm long), white and mauve flowers are clustered at the end of branches. It occurs on lower slopes in this area, and also towards Robertson. (Aug to Oct)

Indigofera brachystachya nenta, beesbossie
A dense, silver-looking, erect shrublet up to 600 mm tall with pinnately divided silver-grey leaves and small rose-pink flowers (*ca.* 7 mm long). It occurs on flats and lower slopes, occasionally on limestone, from the Cape Peninsula to Swellendam. If eaten by sheep it often causes "krimpsiekte". (All year)

Liparia splendens geelkoppie, klipblom
A shrub up to 2,5 m tall with simple oval leaves (*ca.* 40 mm long). The orange flowers are densely clustered into round flowerheads that nod downwards at the ends of branches. It grows on rocky mountain slopes between the Cape Peninsula and Mossel Bay. (All year)

Podalyria cuneifolia keurtjie
An erect, much-branched shrub up to 1,5 m, with silvery-grey leaves (15-20 mm long, *ca.* 1 mm wide) and pink and white flowers. It grows on sandy flats and lower mountain slopes between the Cape Peninsula and George. (Sept to Oct)

Podalyria biflora
An erect, silky shrublet up to 600 mm high with oval leaves (*ca.* 20 mm long) covered with silvery hairs. The attractive pink and white flowers are exceptionally strong-scented. It grows on mountain slopes between the Cape Peninsula and Bredasdorp. (Oct to Dec)

Coelidium ciliare

Indigofera brachystachya

Podalyria biflora

Liparia splendens

Podalyria cuneifolia

❏ **Otholobium fruticans** skaapbostee
An erect or sprawling leafy shrublet up to 1 m, with leaves (*ca.* 10 mm long) divided into three heart-shaped leaflets that are spotted with black glandular dots. The small blue and white flowers (*ca.* 8 mm long) are crowded at the tips of the twigs. It grows in sandy areas between the Cape Peninsula and Port Elizabeth. (June to Jan)

Psoralea aphylla fonteinbos, fonteinertjiebos
An erect to drooping shrub growing to 3 m tall with such small scale-like leaves that the willowy branches appear leafless. The sweet-scented, blue and white flowers (*ca.* 15 mm long) are borne towards the ends of the branches. This plant is characteristic of marshy habitats and streamsides. It occurs from Clanwilliam to Riversdale. (Oct to Feb)

Psoralea pinnata
A small tree up to 3 m high with deeply divided (pinnate) leaves (*ca.* 40 mm long). The blue and white sweetly-scented flowers occur in clusters towards the ends of the branches. It usually grows in damp areas and is distributed from Clanwilliam to Port Elizabeth and further to the interior. (All year)

Rafnia triflora
A robust shrub growing to 2 m, with large, oval leaves (*ca.* 45 mm long) that turn black when dried. Yellow flowers (*ca.* 25 mm long) occur along the branches. It occurs between the Cape Peninsula and Humansdorp. (Sept to Jan)

Sutherlandia frutescens wildegaansie, kankerbos, cancer bush
A shrub up to 900 mm high with greyish, pinnately compound leaves (up to 100 mm long) and striking scarlet flowers (*ca.* 60 mm long). The large, inflated fruit pods are translucent. Isolated plants may be seen along roadsides and it occurs throughout the dry parts of southern Africa. It is especially common in the dune fields of the southern Overberg. This plant has been used as a treatment for cancer, with plants from inland areas allegedly more effective than those growing at the coast. (June to Dec)

Psoralea pinnata

Psoralea aphylla

Otholobium fruticans

Rafnia triflora

Sutherlandia frutescens

GERANIACEAE

Geranium incanum var. incanum bergtee, vrouebossie
A trailing, perennial plant about 250 mm tall with highly divided leaves and pale pink to mauve symmetrically-shaped flowers (*ca.* 25 mm wide) that have petals notched at their tips. It grows in coastal areas from the Cape Peninsula to Port Elizabeth and also northwards in the mountains of tropical Africa. The plant was used to make a "bergtee" (mountain tea) which was used medicinally by early settlers in the Cape. (Aug to Nov)

Monsonia emarginata
A sprawling herbaceous plant up to 250 mm tall with hairy stems and heart-shaped leaves on longish stalks. The symmetrically-shaped, cream flowers are about 25 mm wide. It grows on sandy coastal areas from Bredasdorp to Port Elizabeth, and eastwards to the Transkei. (Sept to April)

Pelargonium betulinum kanferblaar, maagpynbossie
An erect or sprawling shrub up to 1,3 m high. The leaves (10-30 mm long, 7-25 mm wide) are hard and the margins have uneven, red-tipped teeth. The striking flowerheads comprise clusters of one to six flowers that are pink to mauve with darker streaks. The upper two petals are broader and darker coloured than the three lower ones. It occurs on coastal dunes and limestone areas in the southwestern and southern Cape. The vapour from steaming the leaves has been used to treat coughs and other chest problems. (Aug to Oct)

Pelargonium capitatum rose-scented pelargonium
A low-growing, sprawling plant up to 500 mm tall and 1,6 m wide with softly hairy stems and leaves that are sweetly scented when bruised. The velvety, crinkled leaves (*ca.* 45 mm long, 60 mm wide) have three to six shallow to deep lobes and the margins are toothed. The compact flowerheads are borne at the end of longish, upright stalks, and comprise eight to twenty flowers that have pale pink to purple petals with beetroot-purple stripes on the two upper petals. It occurs mainly on coastal dunes often near the shore-line, from Saldanha Bay to Port Elizabeth and also further along the coast to Kwazulu-Natal. (All year)

Monsonia emarginata

Geranium incanum var. *incanum*

Pelargonium betulinum

Pelargonium capitatum

Pelargonium cucullatum wildemalva
A tall shrub up to 2 m high with hooded leaves (*ca.* 45 mm long, 60 mm wide) that are covered with long, soft hairs. The leaf margins are sometimes reddish and irregularly toothed. The flowerheads comprise four to ten large flowers that are pink to purple with dark red veins on the upper two petals. It grows on coastal flats and lower slopes in the area and westwards to the Cape Peninsula. (Sept to Feb)

Pelargonium triste (aand)kaneelbol, rooiwortel
This tuberous species grows up to 600 mm tall and has hairy, dissected leaves (100-450 mm long). The flowerhead of six to twenty flowers is borne at the end of a long, hairy stalk. The flowers, which have yellowish-green to brown-purple petals edged with a lighter margin, emit a musk scent at night. It occurs on sandy or gravelly flats and lower slopes from Clanwilliam to Uniondale and also in Namaqualand. Roots were used to treat diarrhoea and dysentery and also as a vermifuge. (Aug to Feb)

Pelargonium suburbanum subsp. **bipinnatifidum**
A sprawling shrublet up to 300 mm high with stems that are densely covered with woolly hairs and deeply-dissected leaves (*ca.* 40 mm long, 250 mm wide). Flowerheads consist of three to six large, creamy-yellow or pale pink flowers with four or five petals. The upper two petals are much larger than the lower two (three) and have red streaks. It occurs on coastal dune and limestone ridges between Melkbosstrand and Riversdale. (Oct to Jan)

Pelargonium elegans
A tufted shrublet up to 250 mm high. The tough orb-shaped leaves (20-40 mm in diameter) have a coarsely toothed, hairy margin and are borne at the end of long (up to 100 mm) stalks. Flowering branches usually bear two to four elegant flowers that are pale pink to lilac, with dark purple veins on the two upper petals. It occurs on coastal dunes and flats in the area, and westwards to Hermanus. (Sept to Jan)

Pelargonium cucullatum

Pelargonium elegans

Pelargonium suburbanum
subsp. *bipinnatifidum*

Pelargonium triste

OXALIDACEAE

Oxalis purpurea
A low plant up to 150 mm tall with basal, trilobed leaves that have purple under surfaces and hairy margins. The leaves and petals are covered in translucent dots and streaks. Flowers range from brilliant purple or salmon with a yellow throat, to yellow or white. This widespread species is found from Namaqualand to Port Elizabeth. (April to Sept)

Oxalis pes-caprae   geelsuring
Up to 180 mm high, with basal, trilobed leaves consisting of heart-shaped leaflets. The bright yellow flowers occur in loose clusters at the ends of long stems. This is a fairly common plant and occurs from Namaqualand to Humansdorp. (June to Oct)

Oxalis luteola
A dwarf plant up to 80 mm high with basal, trilobed, hairy leaves that are notched at the tips, and bright yellow flowers. It occurs on sandy flats and lower slopes between Clanwilliam and Riversdale. (May to Aug)

Oxalis polyphylla   vingersuring
This plant grows up to 300 mm and differs from the above species in that the leaves have three to seven narrow leaflets folded along their lengths, each bearing two orange warts at their tips. Both the leaves and the rose, lilac or white flowers are clustered at the top of the plant. It occurs between Ceres and Port Elizabeth. (March to June)

Oxalis eckloniana var. sonderi
Up to 100 mm high with basal leaves that are fringed with hairs, and are purple on the reverse. The flowers are mauve or yellow and are borne singly at the end of hairy stalks. It occurs betweem Clanwilliam and Mossel Bay. (April to June)

Oxalis purpurea

Oxalis luteola

Oxalis pes-caprae

Oxalis polyphylla

Oxalis eckloniana var. *sonderi*

ZYGOPHYLLACEAE

Zygophyllum sp. nov.
This newly-discovered species, found on a limestone ridge north of De Hoop Nature Reserve, has not yet been formally described. It has yellow flowers. (May)

Zygophyllum fulvum spekbossie
A straggling shrublet about 500 mm tall with leathery leaves divided into two leaflets and yellow flowers streaked with red that fade to white. It grows on sandy and stony lower slopes from Clanwilliam to Port Elizabeth. (July to Oct)

Zygophyllum morgsana slymbos, skilpadbos
A robust shrub growing to 1,5 m tall. Leaves (*ca.* 250 mm long) are divided into two succulent, oval leaflets and the yellow flowers are borne at the ends of the branches. The large fruits have four prominent papery wings. It occurs in coastal sands from southern Namibia to Port Elizabeth. (June to Nov)

Zygophyllum flexuosum spekbroodbossie
A spreading shrub up to 1 m tall with leathery leaves each subdivided into two leaflets (*ca.* 20 mm long) and bright yellow flowers that have orange-red markings at their centres. It is found on sand dunes and lower slopes from Clanwilliam to Uniondale. (July to Oct)

Zygophyllum sp. nov.

Zygophyllum fulvum

Zygophyllum morgsana (fait)

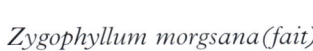

Zygophyllum flexuosum

RUTACEAE

In this family plants have leaves dotted with aromatic oil glands that produce fragrances that are unique to each species.

Coleonema album   aasbossie, Cape may
A dense, finely-branched, fragrant shrub up to 2 m, with spreading branches and needle-like leaves (*ca.* 12 mm long). The solitary, white flowers (7 mm wide) have a dark green disc at the centre, and are crowded towards the ends of branches. It grows on rocky coastal habitats that are subject to strong, salt-laden winds, from near Cape Agulhas to the Cape Peninsula. The aromatic leaves are used by fishermen to remove the odour of red bait (aas) from their hands, hence the common name. (May to Nov)

Acmadenia mundiana
A densely-branched, single-stemmed shrub up to 1,2 m tall with elliptic, shaggy-haired leaves (*ca.* 11 mm long, 4 mm wide). The bright pink flowers (15 mm wide) are closed at the throat. This species is restricted to limestone outcrops between Cape Agulhas and De Hoop Nature Reserve. (April to Oct)

Acmadenia obtusata   duinebuchu
This single-stemmed, highly palatable shrub rarely grows higher than 200 mm. Leaves are linear (10 mm long; 1,5 mm wide) and bright pink flowers (10 mm wide) crowd towards the ends of erect stems. It grows on dunes and limestone hills between Bredasdorp and the Gouritz River and near Port Elizabeth. (April to Nov)

Adenandra gummifera
An upright, single-stemmed shrub up to 1 m high. The leaves (8-15 mm long, 3 mm wide) have oval tips, and the sticky, white to mauve flowers have a pink reverse. Endemic to the Bredasdorp area, it grows on the steep, sandstone south slopes of the Potberg, as well as elsewhere in limestone proteoid fynbos. (Jan to Sept)

Adenandra obtusata   china flower, kommetjieteewater
A robust shrublet up to 1,5 m tall with oval leaves (5 mm long, 3 mm wide). The shiny, white flowers, often with a purple throat, have a red-pink reverse and may be slightly sticky. Restricted to limestone proteoid fynbos in the southern Overberg. (All year)

Adenandra viscida
A low, sparsely-branched shrub up to 500 mm, with leaves that are 10 mm long and 4 mm wide. The sticky, white flowers have a dark pink reverse and occur in clusters at the ends of branches. It is endemic to the southern Overberg where it occurs on rocky, coastal and inland slopes. (Aug to Oct)

Coleonema album

Acmadenia mundiana

Acmadenia obtusata

Adenandra gummifera

Adenandra obtusata

Adenandra viscida

Species of the genus *Agathosma* have flowers which usually occur in flat-topped flowerheads (umbels).

Agathosma cerefolium   anysboegoe, klamboegoe
An aniseed-scented shrub up to 600 mm high with linear-oblong leaves (3-5 mm long). The white, pink or mauve flowers occur in terminal clusters. It occurs on limestone flats and hills between Bredasdorp and Uniondale. (Aug to Jan)

Agathosma collina
A dense shrub up to 1,5 m high with leaves (*ca.* 4 mm long) that are tipped with a rigid hair. Dense clusters of yellow-green flowers occur at the ends of branches. It is common on dunes and limestone hills in the area where it forms conspicuous bright yellow-green patches in the landscape. (Oct to April)

Agathosma riversdalensis
A sprawling shrub up to 600 mm high with leaves about 3 mm long. The white to light-mauve flowers occur in terminal clusters. It is found on sandy limestone flats in the area and further east to Riversdale. (Oct to April)

Agathosma dielsiana
A lax, soft shrub up to 1 m tall. Leaves are about 5 mm long and the terminal clusters of flowers are white or mauve. It grows on coastal dunes and limestone hills in the area and east to George. (April to Oct)

Agathosma serpyllacea
A low shrub with a characteristic licorice aroma and mauve-pink flowers clustered at the ends of branches. It is common on sandy areas near the coast and further inland, as well as on limestone slopes. Distribution is between Clanwilliam and Mossel Bay. (Jan to Dec)

Agathosma riversdalensis

Agathosma collina

Agathosma dielsiana

Agethosma serpyllacea

Agathosma cerefolium

The generic name *Euchaetis* means "beautifully tufted with hairs", referring to the bearded inner margin of the petals.

Euchaetis longibracteata
An erect, single-stemmed shrub up to 800 mm, with branches that are partially leafless showing smooth, brown bark. The oblong leaves (9 mm long, 3 mm wide) curve upwards and have scattered spiky hairs. The whitish-pink flowers occur at the ends of branches and are surrounded by a distinctive rosette of long, pale involucral bracts. It is endemic to this area where it grows on limestone proteoid fynbos. (Dec to April)

Euchaetis meridionalis
A compact shrub up to 1,5 m high with leaves that give off a resinous exudate at their point of attachment to the stem. The flowers are pale pink. It is endemic to limestone proteoid fynbos in the eastern part of the southern Overberg. (April to Dec)

Euchaetis burchellii
This shrub, about 1 m tall, has a persistent rootstock enabling it to resprout after fire. The leaves are narrow (4,5 mm long, 1,5 mm wide) and the white or pink flowers occur at the ends of branches. It grows on coastal sands and limestone between Bredasdorp and George. (All year)

Members of the genus *Diosma* have small flowers (less than 10 mm wide) with a wavy, wax-like disc in the centre, a short style and stamens that do not protrude beyond the petals.

Diosma subulata
An erect, single-stemmed shrub growing to 1.8 m, with a dense crown of white flowering heads. The leaves are lance-shaped (*ca.* 17 mm long. 2 mm broad) ending in a sharp point, and the margins are hairy. It occurs in sandy soils (neutral sand proteoid fynbos) in a small area between Stanford and Pearly Beach, as well as at Hawston. (March to Aug)

Diosma guthriei
A spreading, rigid shrub up to 400 mm, with a persistent rootstock. The oval leaves (3-6 mm long, *ca.* 3 mm wide) have thick margins and a blunt callus at the tip. Clusters of two to three white flowers occur at the ends of branches. It is endemic to limestone areas between Pearly Beach and Arniston. (July to Aug)

❏ Diosma haelkraalensis
A low, sprawling shrub restricted to exposed limestone rocks on the farm Groot Hagelkraal near Pearly Beach. The small, recurved leaves have translucent margins. (April to May)

Diosma guthriei

Diosma haelkraalensis

Diosma subulata

Euchaetis burchellii

Euchaetis longibracteata

Euchaetis meridionalis

POLYGALACEAE

Polygala myrtifolia
Augustusbossie, blouertjieboom

A tall, lax shrub up to 2,5 m tall with leaves 20-30 mm long and 4-8 mm wide. The purple flowers have a keeled and fringed front petal and are superficially pea-like. Flowering is most prolific during spring but also occurs during the rest of the year. (All year)

Polygala umbellata
A low, erect or spreading shrub up to 300 mm high. It has magenta pea-like flowers, and occurs on dry, sandstone and limestone slopes at low altitudes between Ceres and Riverdale. (Aug to Oct)

Nylandtia spinosa
skilpadbessie

A spiny, shrub growing up to 2 m high and sparsely covered with small, narrow leaves (3-6 mm long). The numerous pink to purple flowers produce round, red, fleshy fruits during the summer. These are widely eaten by tortoises and other animals. (April to Oct)

Muraltia collina
An erect shrub up to 350 mm high with bunches of spine-tipped leaves (5-8 mm long) and small, purple flowers (*ca.* 5 mm long). It occurs on the lower sandstone slopes in the area, for example Potberg, as well as in the vicinity of Swellendam. (Sept to Jan)

Muraltia satureoides var. satureoides
An erect to spreading shrub up to 600 mm high with hairy young stems. It has crowded bunches of oblong leaves (3-8 mm long) that are spine-tipped when young and white or pink flowers (*ca.* 5 mm long). This species grows in dune fynbos areas between the Cape Peninsula and Knysna. (Aug to May).

Nylandtia spinosa (fruit and flowers)

Polygala myrtifolia

Polygala umbellata

Muraltia collina

Muraltia satureioides var. *satureioides*

EUPHORBIACEAE

Clutia ericoides
A wiry shrublet with small, slightly fleshy, shiny leaves and inconspicuous, yellow-green cup-shaped flowers along the stem. Male and female flowers occur on separate plants. It occurs between Piketberg and Port Elizabeth. (May to July)

Euphorbia erythrina   pisgoed(bossie)
An erect plant up to 600 mm tall with stems that are covered with small, oval leaves (*ca.* 8 mm long) and copious milky juice. The unusual-looking green-yellow "flowers" of the genus *Euphorbia* are composite flowerheads with a central, stalked female flower that has three 2-lobed styles and an outer ring of small male flowers, the whole surrounded by a cup of flaring bracts. Loose clusters of these flowerheads are borne at the ends of the stems. It occurs on flats and slopes in the Cape Peninsula, Bredasdorp and Riversdale areas. (June to Oct)

Euphorbia tuberosa
A herbaceous plant about 100 mm high with a tuberous root and a tuft of strap-shaped leaves at ground level. Loose clusters of yellow-green cup-shaped flowers are borne on lax, longish stems. The plants have copious, milky juice. It occurs in this area and also westwards to the Cape Peninsula and Vanrhynsdorp. (April to Sept)

Clutia ericoides   *Euphorbia erythrina*

Euphorbia tuberosa

ANACARDIACEAE
Cape members of the genus *Rhus* characteristically have compound leaves comprising three leaflets.

Rhus lucida
slaptaaibos, kraaibessie

A sprawling, much-branched shrub up to 3 m tall with olive-green leaves that have a varnished appearance. The yellow flowers are clustered and the round fruits are a shiny, dark brown. It occurs in coastal bush and on inland slopes from the Cape Peninsula to Port Elizabeth, as well as in the Little Karoo. (May to Sept)

Rhus rosmarinifolia
This straggling shrublet (usually less than 1 m tall) has straight to curved, narrow leaflets that are wrinkled and greyish green above and white below. The cream flowers are borne in panicles and the oblong fruit are woolly. It grows in rocky soils in fynbos between Clanwilliam and Port Elizabeth.(April to Sept)

Rhus crenata
rosyntjiebos

A much-branched shrub growing to 4 m. The trifoliate leaves are dark green above and slightly paler below and the broadened ends are scalloped. The white flowers are borne in loose clusters and the round, shiny fruit are bluish to dark brown. This widespread species grows in dune thickets between the Cape Peninsula and southern Natal. (April)

Rhus glauca
kroestaaibos

A much-branched shrub up to 4 m high. The trifoliate leaves are covered with a green-blue powdery bloom and the greenish-white flowers occur in loose clusters. The oblong fruits are a shiny chestnut-brown. It grows in coastal and inland thickets from Piketberg to the Eastern Cape.(Jan to Mar)

Rhus crenata

Rhus glauca *Rhus lucida*

Rhus rosmarinifolia

CELASTRACEAE

Cassine peragua lepelhout
A large shrub or tree, 3-10 m tall, with leathery, almost circular leaves (70-100 mm long) that have irregularly serrated margins. There are loose clusters of small, cream flowers that produce purplish-black, fleshy fruit. It is usually multi-stemmed and has a characteristic orange bark. A tree of both the forest and coastal thickets, it occurs between the Cape Peninsula and Port Elizabeth and also northwards to Mpumalanga. (Jan to June)

Cassine maritima
A low, rigid dense shrub about 900 mm tall with spreading branches and angular green twigs. The leaves (*ca.* 70 mm long) are thick and slightly fleshy and the small flowers are whitish. This species grows in dune thickets and fynbos between the Cape Peninsula and Port Elizabeth. (April to Jan)

Maytenus procumbens
A scrambling shrub or small tree, 1-6 m tall. The leathery leaves (*ca.* 70 mm long) have hardened margins with three to five spine-tipped teeth that are conspicuously bent backwards. The small, greenish-white flowers occur in clusters and the capsular fruit are yellow-orange, splitting to release the orange seeds. This species grows in coastal thickets and forests from Bredasdorp area to tropical Africa. (April to July)

Pterocelastrus tricuspidatus kers(ie)hout, cherrywood
A shrub or small tree 4-7 m tall. The oval leaves (*ca.* 50 mm long) are thick, leathery and shiny above and a paler green below. They have a shallow-notched apex and the margins are slightly rolled under. Young leaves are faintly veined and leaf stalks are pink-red. The small, fragrant flowers are yellowish to creamy-white and occur in compact clusters. The orange-yellow capsular fruits have three lobes, each characteristically with one or two winged protuberances. It occurs in dune thickets and mountain forests between the Cape Peninsula and Port Elizabeth, as well as northwards to Mpumalanga. (April to July)

Pterocelastrus tricuspidatus
(fruit shown below)

Cassine maritima

Cassine peragua

Maytenus procumbens

RHAMNACEAE

The genus *Phylica* is confined to southern Africa, as well as to a few islands in the South Atlantic and Indian Oceans. It comprises shrubs with tough leaves that are rolled downwards to cover much of the lower surface.

Phylica stipularis   hondegesiggie

A much-branched shrub up to 900 mm high with slender, wiry branches that are covered with white fur. The leaves (10-20 mm long) have a heart-shaped base and closely rolled margins and there are two 3 mm long, brown, linear stipules where the leaf stalk joins the stem. It occurs on sandy flats and lower slopes between Clanwilliam and Knysna. (May to Sept)

Phylica selaginoides

A wiry, branching shrub up to 400 mm high, with closely set, inward-rolled, linear leaves, and white flowers. Endemic to limestone proteoid fynbos of the southern Overberg. (Aug to Sept)

Phylica purpurea

A robust shrub about 800 mm tall with hairy stems. The leaves (8-10 mm long) are hairless above and have very hairy undersides. Flowers are woolly and dirty-white on the outside, with the inside yellow and later turning to pink and reddish-brown. It grows in acid sand proteoid fynbos in the southern Overberg and also northwards to the Swartberg and beyond to Humansdorp. (Mar to Oct)

Phylica pubescens var. orientalis   veerkoppie

This attractive shrub grows to 1,5 m and has slightly hairy, ericoid leaves (10-20 mm long). The flowers occur in feathery, greenish-yellow heads that are about 40 mm wide. It occurs between the Cape Peninsula and Riversdale. (May to Aug)

Phylica dodii

This species is about 500 mm high and has wiry, reddish branches bearing closely-set linear leaves (8-20 mm long) and whitish flowers. It occurs on sandy habitats in the area and also west to the Cape Peninsula. (June to Sept)

Phylica ericoides

A shrub up to 1,5 m high with narrow, ericoid leaves about 8 mm long and dense clusters of small, whitish flowers. It grows on dunes and lower slopes from the Cape Peninsula to Port Elizabeth and beyond. It is widely harvested as a "green". (All year)

Phylica purpurea

Phylica selaginoides

Phylica ericoides

Phylica dodii

Phylica pubescens var. *orientalis*

Phylica stipularis

MALVACEAE

Hibiscus trionum
An erect or straggling plant up to 1 m tall with hairy stems. The leaves are lobed and coarsely toothed and the yellow to cream, dark-centred flowers (*ca.* 45 mm wide) have an inflated calyx. This cosmopolitan weed grows in fynbos areas between Clanwilliam and Port Elizabeth as well as elsewhere in South Africa. (Sept to Feb)

Anisodontea scabrosa
An erect, much branched shrub up to 2 m tall with coarsely toothed, rough leaves and pink flowers with spreading petals. It is widespread in coastal areas from Saldahna to Port Elizabeth as well as elsewhere in southern Africa. (All year)

STERCULIACEAE

Hermannia concinnifolia   poprosies, geneesbossie
A shrublet up to 900 mm with closely-spaced, small leaves (*ca.* 10 mm long). The dainty yellow flowers have a red calyx and the petals are twisted into a tiny pin wheel or "doll's rose" (poprosie). It is restricted to limestone from the Stanford area to Riversdale. (Aug to Oct)

Hermannia ternifolia   tandebossie
A prostrate or sprawling shrublet with small, slightly toothed, grey, velvety leaves. The small, pin-wheel flowers are yellow tinged with orange-red. It is a plant of sand dunes and coastal limestone from Saldanha to the Bredasdorp area. (Sept to Nov)

Hermannia trifoliata
A sprawling shrublet about 400 mm high with grey, velvety, reflexed leaves that are densely arranged along the stem. The orange to red pin-wheel flowers have an inflated calyx. This species is confined to limestone in this area and east to Riversdale. (Aug to Sept)

Hibiscus trionum

Anisodontea scabrosa

Hermannia ternifolia

Hermannia corcinnifolia

Hermannia trifoliata

VIOLACEAE

Viola decumbens wild violet
A perenial, often woody shrublet that may form cushion-like bushes 80-250 mm tall. It has linear leaves (15-50 mm long, 0,5-2 mm wide) and faintly-scented, blue-violet flowers that have a tubular spur. It grows in mountainous fynbos areas from the Cape Peninsula to Bredasdorp. (July to Dec)

PENAEACEAE

Brachysiphon acutus
An erect shrublet up to 400 mm tall. The upper parts of the branches have four prominent ridges and the oval, sessile leaves (6-15 mm long) have pointed tips. The pink to purplish flowers (*ca.* 10 mm long) are tubular and are clustered at the ends of branches. It grows on the mountains of the area and also at Hermanus and Caledon. (Oct to Dec)

Penaea mucronata
An erect shrub up to 1 m tall with closely over-lapping leaves (*ca.* 8 mm long) that have pointed tips. The dumpy, tubular, yellow flowers occur in clumps at the tips of branches. It grows on acid sandy flats and slopes in the area, as well as further west to the Cape Peninsula and north to the Langeberg. (All year)

Saltera sarcocolla vlieëbos(sie)
An erect shrub up to 1 m tall with branches that are leafy in the upper parts and bare and knotty below. The oval leaves overlap each other to encircle the stem. The large, tubular flowers (40 mm long) are rose to purple and are grouped at the end of branches. It grows in moist fynbos on slopes and flats in the area as well as westwards to the Cape Peninsula. (All year)

Viola decumbens

Brachysiphon acutus

Penaea mucronata

Saltera sarcocolla

THYMELAEACEAE

Plants in this family have stems with very strong fibres - when branches are broken the bark forms long, tough strips.

Gnidia vesciculosa   naeltjies
A robust shrub about 200 mm tall with terminal clusters of small, tubular white flowers. It is endemic to this area where it grows on marshy flats and slopes. ((July to Oct)

Gnidia viridis
Plants are about 150 -300 mm tall and have a persistent rootstock enabling them to resprout after fire. Leaves are 8-10 mm long and 2-3 mm wide and the terminal clusters of small, yellow, tubular flowers are circled by a whorl of bract-like leaves. It occurs in gravelly and deep acid sands in the area. (Oct to Dec)

Gnidia albicans
A lax, silvery shrub up to 2 m tall with small, hairy leaves. The small, creamy-yellow flowers have a well developed tube and they are clustered at the tips of branches. It is found on sandstone slopes from the Cape Peninsula to Bredasdorp. (May to Nov)

Gnidia pinifolia
An erect shrub growing to 1 m tall with small, linear leaves (*ca*. 10 mm long). The long-tubed white flowers (*ca*. 12 mm long) are densely crowded at the ends of branches. This species occurs on sandy flats and mountains between Piketberg and Port Elizabeth, and also to the interior. (All year)

Struthiola argentea   roemenaggie, aandgonna
A shrublet about 600 mm tall with small, oval leathery leaves that are curved downwards. The small, long-tubed, yellowish flowers are borne along the upper sections of branches, contrasting with *Gnidia* species where flowers are clustered at the ends of branches. This is a species found on coastal flats and lower slopes between Stellenbosch and Port Elizabeth. (May to Dec)

Gnidia vesiculosa

Gnidia viridis

Gnidia pinifolia

Gnidia albicans

Struthiola argentea

Lachnaea aurea
A slender shrub up to 1 m with small leaves and chrome-yellow flowers densely clustered into terminal heads. It is confined to the Bredasdorp area where it grows in Elim and acid sand proteoid fynbos of the coastal flats and lower slopes. (June to Sept)

Passerina rigida gonnabas
A robust shrub up to 2 m tall with tiny, ericoid leaves. The numerous, small flowers are yellow to red and the orange fruits ripen in summer. It is a pioneer species on coastal sand dunes from the Cape Peninsula to Kwazulu-Natal. The berries are used for dye. (Sept to Nov)

Passerina ericoides skilpadbesssie
A shrub up to 1 m tall with drooping branches and ericoid leaves. Large numbers of greenish or yellow flowers occur along the branches. They are pollinated by wind, and the fruit is a scarlet berry that is evident in the summer months. It grows in dune fynbos from the Cape Peninsula to the Bredasdorp area. (Oct to Nov)

ARALIACEAE

Cussonia thyrsiflora kiepersol
A scrambling shrub or tree growing to 2-5 m high. The large, round, palmate leaves are crowded near the ends of the branches and it has a sturdy spike-like arrangement of small, greenish-yellow flowers. It grows in coastal thicket from the Cape Peninsula to Port Elizabeth. (Nov to Dec)

Passerina ericoides

Lachnaea aurea

Passerina rigida

Cussonia thyrsiflora

APIACEAE

Centella virgata
A sprawling, lax shrublet up to 500 mm high. The leaves are linear (up to 15 cm long) and the small, yellow to greenish flowers are borne in umbels. It occurs on flats and lower slopes from Clanwilliam to Port Elizabeth. (June to Jan)

Anginon difforme (= *Rhyticarpus difformis*)
A rigid shrub growing to be 1-2 m tall with woody, jointed stems and few erect branches. Clusters of narrow, linear leaves (150 mm long, 3 mm wide) occur at the nodes. Small greenish-yellow flowers are borne in umbels. This species grows on lower mountain slopes between Tulbagh and Port Elizabeth. (Feb to April)

Arctopus echinatus   sieketroos, platdoring
A stemless, perennial plant with a substantial underground tuber. The large, shiny, leathery leaves (*ca.* 100 mm long) are pressed to the ground and have scattered spines. White to pinkish flowers are borne in dense umbels on separate male and female plants. The spiny fruits are dispersed by animals, including humans. It occurs on flats and lower slopes from Nieuwoudtville to Port Elizabeth. (May to Aug)

Peucedanum galbanum   bergseldery, blisterbush
A robust shrub between 1-3 m tall with compound leaves (40-50 mm long). The small, yellow flowers are borne in an umbel. ***Avoid contact with this plant***, especially in hot weather, as this causes a strong skin irritation resulting in itching and blisters a day or two later. It occurs on mid to upper slopes between Piketberg and Riversdale. (July to Feb)

Dasispermum suffruticosum
A spreading, perennial plant 200-500 mm tall with jointed, slightly fleshy stems growing along the ground. The cream flowers are arranged in dense umbels. This is a plant of coastal dunes and occurs from Saldanha Bay to Port Elizabeth and further east to Kwazulu-Natal. (Aug to Mar)

Centella virgata

Peucedanum galbanum

Arctopus echinatus
(Below ♀, below right ♂)

Dasispermum suffruticosum

Anginon difforme

ERICACEAE
(Note: These minor genera (i.e. non *Erica*) will be included under *Erica* when the research has been published)

Anomalanthus scoparius
A low, spreading shrublet up to 250 mm high, with deep pink flowers (3 mm long, 1 mm wide) with dark brown, protruding anthers, the flowers arranged along the main branches. The calyx is enlarged and protects the tiny fruits. It is very common from Grabouw to George and inland to the Little Karoo mountains. (Feb to May)

Erica albertyniae
A low compact shrublet to 300 mm high with pink flowers having black exserted anthers and arranged in small heads at the ends of branches. The species occurs on lateritic or limestone flats south of Heuningrug and eastwards towards Potberg. (Feb to June)

Simocheilus purpureus
A compact or sparse erect shrublet up to 500 mm high with terminal heads of pink flowers with exserted dark anthers. It occurs on sandy flats and lower slopes from the Cape Peninsula to Potberg. (Mar to Oct)

Syndesmanthus articulatus
A compact shrublet up to 300 mm high with terminal clusters of hairy white pink flowers with exserted dark brown anthers. It occurs on sandy flats and lower slopes from the Cape Peninsula to Mossel Bay. (Mar to Oct)

Thoracosperma puberulum
A rounded shrub up to 500 mm high with dusky pink to maroon flowers. It occurs on dry lower slopes between Hermanus and Riversdale and is especially common in Elim fynbos. (July to Oct)

Simocheilus purpureus

Anomalanthus scoparius

Syndesmanthus articulatus

Simocheilus albertyniae

Thoracosperma puberulum

Erica is the largest genus in the fynbos with 426 species in the south western Cape. There are 106 species recorded in the southern Overberg and 39 are illustrated below.

Erica ampullacea   bottelheide, bottle/sissie heath
An upright, sparsely branched, lax shrub up to 600 mm tall. The striking pink-white flowers (18-24 mm long) are clustered in groups of three to four at the end of branches, and have the appearance of polished china flasks. Occasional plants can be found on the Heuningberg, the Pearly Beach area and in the vicinity of Elim. (July to Dec)

Erica berzelioides
An erect, rounded shrub up to 600 mm high. The tubular, slightly curved flowers (*ca.* 30 mm long) occur in clusters of three at the tips of branches and are deep pink at the base turning to white at the upper end. It is locally frequent in coastal sandy soils. (April to June)

Erica bodkinii
An erect shrub up to 600 mm high, with soft, very finely hairy flowers (*ca.* 10 mm long) hanging in clusters of one to three and changing from white to pink, red and then brown. It grows on the cool, moist slopes of the Bredasdorp and Napier mountains, forming a beautiful display in winter. (June to July)

Erica bruniades
A diffuse, slender shrub up to 450 mm high with pale pink to rosy, hairy flowers with protruding anthers. This widespread species occurs near streams or seeps in the area, as well as further west to the Cape Peninsula and north to the Gifberg. (July to Jan)

Erica bruniifolia
Sprawling bushes up to 450 mm high with tightly packed creamy white flowerheads that hang head downwards. Brown stamens protrude from the small (*ca.* 4 mm long) flowers. This species grows on the sandy flats and lower slopes in the Bredasdorp area and in the De Hoop Nature Reserve. (July to Oct)

Erica calcareophila
A shrub up to 250 mm high that is found only on limestone ledges and cliffs in the Pearly Beach area at Groot Hagelkraal. The urn-shaped, waxy white flowers (*ca.* 10 mm long) are borne in clusters of up to six forming a beautiful show when in bloom. (July to Sept)

Erica ampullacea

Erica bruniifolia

Erica berzelioides
Erica calcareophila

Erica bruniades
Erica bodkinii

Erica casta
A tall, straggly plant up to 1.2 m high with sticky, white, translucent flowers (10-14 mm long) near the tips of the slender branches. It occurs on clay banks in the Viljoenshof area in one of the small, remnant patches of renosterveld. (Oct to June)

Erica cerinthoides rooihartjie
This well-known species has a persistent rootstock enabling it to resprout after fire when it is characteristically seen in bloom. The bright red, tubular flowers (25-35 mm long) are hairy and sticky, and are arranged in tightly packed clusters. It has a wide distribution and besides occurring throughout fynbos areas in the Cape, it is also present throughout many other areas of the eastern part of southern Africa. (All year)

Erica coccinea vlakteheide/heath
A rigid, stoutly branched shrub up to 1,2 m with small bunches of leaves. The flowers (6-17 mm long) are variable and may be smooth or hairy, dry or sticky, and have a wide range in colour (red, pink, orange, brown, yellow and green). A characteristic feature is the protruding, brown anthers which hang down from the slightly inflated flower tube. It occurs from Clanwilliam to Knysna in habitats ranging from dry sandy flats to wet mountainous areas. (All year)

Erica colorans tregterheide(heath)
An erect, sparsely branching shrub up to 1,8 m high. It has clusters of flowers (*ca*. 15 mm long) at the end of densely leafed branches. They are white with a dark pink tip, and the tube constricts at its mouth (this feature distinguishes it from *Erica perspicua*, a similar looking species which occurs in the Betty's Bay - Hermanus area). It grows in marshes and vleis of the area. (Aug to Oct)

Erica corifolia
A slender shrub about 450 mm tall with terminal clusters of small, pink flowers (3-10 mm long). The urn-shaped corolla turns brown at its tip soon after opening. It grows on sandy flats and plateaux from the Cape Peninsula to Uniondale. (Oct to May)

Erica discolor
A large, spreading resprouter up to 1.8 m high. The red to pink flowers (18-24 mm long) have lighter white, yellow or green tips. It grows on coastal and lower mountain slopes from Betty's Bay to Port Elizabeth. (All year)

Erica coccinea

Erica colorans

Erica cerinthoides

Erica corifolia

Erica discolor

Erica casta

Erica filipendula   paasfeesblom
An erect shrub up to 600 mm high. The pendulous flowers have thread-like flower stalks, and show a wide range of shapes and colours. The most common forms have mauve, pink, red or yellow egg-shaped flowers (*ca.* 9 mm long), while other forms have white to green tubular flowers (10-18 mm long). Plants occur on sandy flats as well as in marshes and are restricted to the Viljoenshof, Soetendalsvlei and Soetanysberg areas of the southern Overberg. (April to July)

Erica globulifera
This species shares the same habitats and distribution as *Erica filipendula* and was previously considered to be a variety of it. However, its mauve to pink-white flowers (*ca.* 5 mm long) are more prolific, and are smaller, rounder and usually paler than the latter. (April to June)

Erica grisbrookii
A sturdy shrub up to 900 mm high with stout, erect branches and greyish-green leaves. The flowering branches are bare in the lower half and the hard, waxy creamy to green-white flowers (11 mm long) are crowded in the upper section. Found on mid-altitude rocky slopes and plateaux between Stanford and Bredasdorp. (June to July)

Erica imbricata   kêr-kêr
An erect shrub up to 800 mm high with small bunches of leaves. The white or pink flowers (*ca.* 3 mm long) have protruding brown anthers. It grows on dry, sandy flats and lower slopes from the Gifberg to Port Elizabeth. (Feb to Nov)

Erica irbyana   bottelheide/heath
A sparse shrublet up to 500 mm high with slender branches bearing pink to deep red, inflated, sticky flowers (8-14 mm). It is common on lower mountain slopes from Hawston to Elim. (Oct to March)

Erica filipendula

Erica globulifera

Erica grisbrookii

Erica imbricata

Erica irbyana

Erica irregularis
An erect, sturdy shrub growing up to 1,2 m tall with upright, lax branches that are covered with pale pink, rounded flowers (*ca.* 5 mm long) that have a constriction at the mouth, and are borne on fairly long, woolly stalks. It grows in sandy habitats between Stanford and Gansbaai, and in late winter profuse flowering paints the landscape pink, especially in the Grootbos and Die Kelders area. (June to Sept)

Erica lineata
An erect, well-branched shrub up to 600 mm high with long, soft leaves giving it a feathery appearance. The pendulous, inflated tube-shaped flowers (7-10 mm long) vary from white to red, and have protruding, brown anthers. The species occurs on neutral sand proteoid fynbos near Bredasdorp, Soetanysberg, Cape Agulhas and Pearly Beach. (Oct to July)

Erica longiaristata   klokkie
A slender, lax shrub up to 300 mm high with small, urn-shaped flowers (*ca.* 4mm long) varying from white to rosy pink. The long awns on the anthers, which have given the plant its name, can only be seen after dissecting the flower tube. This species has a narrow distribution, found only in sandy flats and low slopes near Baardskeerdersbos, Elim, Napier and Soetanysberg. (Nov to March)

Erica longifolia
A lowish, compact bush to a tall willowy plant up to 900 mm high. The leaves are usually long, rigid and sharp-tipped, but they may also be quite short. The fairly tubular flowers (12-22 mm long) are slightly constricted at the mouth and are bunched together at the end of the branches. They range widely in colour from white, yellow, orange, brown, pink, red and purple, as well as being two-toned. It is a common species on sandy mountain slopes between Paarl and Bredasdorp. (All year)

Erica mariae
A tall, woody, lanky plant that may grow to 2 m in height. The waxy and shiny, tubular, dark red flowers (*ca.* 24 mm long) are fairly densely arranged at the end of branches. It occurs on limestone areas from the Heuningrug to Still Bay and also on the acid soils of Potberg. (Jan to March)

Erica irregularis

Erica lineata

Erica longiaristata

Erica mariae

Erica longifolia

Erica multumbellifera
This low, bushy shrub is about 400 mm tall and forms a bright splash in the landscape during its flowering peak when it is covered with numerous purple-red flowers. The small, round flowers occur in terminal, drooping umbels. It occurs on sandy flats and lower slopes from the Cape Peninsula to Riversdale. (Nov to May)

Erica nudiflora
An erect shrub up to 300 mm tall. A feature of this plant is that it has hairy leaves and stems, but the flowers are smooth. The tubular flowers (*ca.* 4 mm long) range from pale pink to bright red and have protruding brown anthers. This is a common, widespread species that occurs on dry, stony lower slopes in the Bredasdorp area as well as to the Cape Peninsula and Clanwilliam. (Feb to April)

❏ Erica oblongiflora   groentaaiheide
An erect shrub up to 1m tall with rigid, spreading branches. The green-yellow, sticky flowers (*ca.* 8 mm long) occur in umbels (bunches of flowers with equal stalk lengths) and are oblong to urn shaped. This plant is restricted to limestone hills south-west of Bredasdorp. (April to July)

❏ Erica occulta
This compact, densely-leafed plant (*ca.* 300 mm tall) does not initially appear to be an *Erica*. However, closer inspection reveals the small, yellowish, hidden flowers (*ca.* 8 mm long). This extremely rare and unusual species is restricted to small populations on steep limestone cliffs at Groot Hagelkraal near Pearly Beach. (Aug to Oct)

Erica placentiflora   kêr kêr, tjêr-tjêr, klokkiesheide
An erect shrub up to 900 mm with long branches bearing numerous pink-purple, urn-shaped flowers (*ca.* 5 mm long) with protruding black stamens. This species is very similar to *E. imbricata*. It occurs in dry, sandy habitats in the Bredasdorp to Swellendam area, as well as between Clanwilliam and Stellenbosch, and in the Swartberg mountains. (July to Nov)

Erica multumbellifera

Erica oblongiflora

Erica occulta

Erica nudiflora

Erica placentiflora

Erica plukenetii   snotbel, hangertjie
An erect shrub up to 1 m tall with feathery leaves and numerous tubular, pendulous flowers (13-18 mm) from which brown anthers protrude. The colour varies from white, yellow and green to pink and deep red. This species is common in sandy soils of this area and is also widespread elsewhere in Cape fynbos areas. (All year)

Erica propinqua
An erect, lanky shrub up to 900 mm high with deep pink, urn-shaped flowers (*ca.* 5 mm long) that are borne in clusters of three at the ends of branches and tend to hang down on relatively long stalks. It is restricted to limestone from Die Kelders to De Hoop Nature Reserve. (Aug to Oct)

❏ Erica regia   Elim heath, Elimsheide
A striking, erect shrub up to 700 mm tall with shiny, tubular flowers (14-18 mm) that range from being uniform red, to white with a red tip (the "Elim heath") and occasionally a green interface. The species grows in sandy soils between Elim and Bredasdorp. (April to Oct)

Erica rhopalantha
A compact, rounded shrub about 300 mm high with many, dark pink to mauve urn-shaped flowers (*ca.* 4 mm long) at the ends of slender branches. It is common in sandy habitats in the Ratel River and Soetanysberg areas and also westwards to Betty's Bay. (Nov to May)

Erica scytophylla
An erect shrub up to 1,5 m high with thick, closely packed leaves and white to pink-purple, bell-shaped flowers. It grows on limestone hills and dune sands between Bredasdorp and Cape Infanta. (Sept to Dec)

Erica sessiliflora   green heath
This species is usually about 1 m tall and has short, dense spikes of tubular, green flowers (16-30 mm long). These spikes mature to form reddish, lumpy heads of fruit that remain on the plant for many years. This unusual characteristic makes the species easy to identify. It is widespread in marshy areas of Cape fynbos. (April to Sept)

Erica propinqua

Erica plukenetii

Erica regia

Erica sessiliflora

Erica scytophylla

Erica rhopalantha

❏ **Erica shannonea** bottelheide, bottle heath
A rounded shrub up to 450 mm high with clusters of 8-10 shiny, white flowers (*ca.* 25 mm long) at the tips of branches. The flowers have a constricted throat flaring out into star-shaped lobes with four points. This rare and spectacular species grows in stony, dry habitats on lower slopes between Stanford and Bredasdorp. (Dec to Jan)

Erica spectabilis
A sturdy shrub up to 700 mm high bearing numerous umbels of white flowers (2-5 mm long). It is found in coastal areas between Bredasdorp and Still Bay, as well as further inland on dry mountain ranges beyond the Overberg. (June to Dec)

Erica tenella
An upright shrub up to 1 m high with bright to dark pink, urn-shaped flowers (*ca.* 5 mm long) that are densely arranged at the ends of lax flowering branches. This species grows on the cool slopes of mountains in the area, often forming dense stands. (All year)

Erica vernicosa
Prostrate mat-forming shrublets with sticky mauve flowers and dark exserted anthers. It grows on limestone flats only in the De Hoop area. (Mar to June)

Erica vestita trilheide/heath, wide mouth heath
A compact shrub about 900 mm tall with densely packed, thin leaves that shimmer in the breeze. The tubular flowers (17-25 mm) range from white through pink to red and are sometimes two-toned pink and white. It occurs in both dry and moist habitats on the lower slopes of mountains between Worcester and George. (Aug to May)

Erica vogelpoelii
An erect shrub up to 500 mm high with slender branches and deep pink-purple, flask-shaped flowers (7-9 mm long) that end in spreading, pointed lobes. It is restricted to sandy, moist slopes on the Bredasdorp-Napier mountains where flowering plants form a bright splash in late summer. (Dec to April)

Erica shannonea

Erica spectabilis

Erica vernicosa

Erica tenella

Erica vestita

Erica vogelpoelii

MYRSINACEAE

Myrsine africana (wilde)mirting, vlieëbos, Cape myrtle
An erect, shrub about 1,5 m high with small leaves (5-20 mm long) that have finely serrated margins. It has small, pinkish flowers and the globular fruits are blue-black. Found in forest and in forest margins from the Cape Peninsula to Port Elizabeth, as well in tropical Africa and Asia. (Oct to May)

PLUMBAGINACEAE

Limonium scabrum var. **scabrum** sea lavender, statice
A branched, tufted shrublet up to 250 mm tall with a basal rosette of leaves. The papery, lavender to violet flowers occur in a highly-branched flowerhead. It grows in dune and limestone fynbos as well as along rocky shores and estuaries from the Cape Peninsula to Port Elizabeth. (Oct to May)

Limonium anthericoides brakblommetjie
A tufted, perennial up to 500 mm high, with slender branches and papery, white to pink-lilac flowers. It occurs in vlei and marshy areas from Danger Point to Cape Infanta. (Dec to Feb)

SAPOTACEAE

Sideroxylon inerme melkhoutbos, melkhout boom, milkbush
A densely leafy shrub or rounded tree growing to about 10 m tall with oval, shiny green leaves (*ca.* 100 mm x 40 mm) and stems that bleed a milky latex. The flowers are small and greenish-white and the round, fleshy fruits are purple-black. Dense thickets of large trees are a characteristic feature of the southern Overberg. However this species can also occur as a stunted shrub on limestone or wind-blown rocky coasts. It is a dominant tree in coastal thicket and forest between the Cape Peninsula and Port Elizabeth and is also widely distributed along the coast to Tanzania. The wood was used to make wagons. It is now a protected species. (Dec to June)

Limonium anthericoides

Limonium var. *scabrum*

Myrsine africana   *Sideroxylon inerme*

EBENACEAE

Euclea racemosa seaguarri
A shrub or tree 1-6 m tall. It has leathery, oval leaves that are paler green below than above and are rolled under at the margins. The small creamy-white flowers occur in short spikes (*ca.* 40 mm long) and mature to form round, thinly-fleshed, black fruits. It occurs in coastal dune thickets and forests from Lamberts Bay to East London and also in Namaqualand. (Dec to June)

OLEACEAE

Olea capensis subsp. **capensis** ysterhout, wilde-olyfboom
A shrub or small tree up to 10 m tall with opposite, oblong leaves (*ca.* 50 mm long) that have thickened, usually wavy margins. The small, whitish-cream flowers are sweetly scented and occur in clusters along the sides of, or at the ends of branches. The round, fleshy fruits become purple when mature. It is found in thicket dunes and lower slopes between Clanwilliam and Port Elizabeth and northwards to tropical Africa. (Sept to Mar)

Olea exasperata slanghout
This shrub or small tree grows to 2 m tall and has leathery, linear-oblong, opposite leaves (*ca.* 70 mm long) with tips that are usually bent backwards. The small, scented, whitish-cream flowers are clustered into heads at the ends of branches and the thinly-fleshed, round fruits are yellowish-purple when ripe. It grows in dune thickets from the Cape Peninsula to Port Elizabeth. The root was used as an antidote to snake-bite. (Aug to Oct)

Olea exasperata

Olea capensis subsp. *capensis*

Euclea racemosa

GENTIANACEAE

Chironia baccifera
aambeibossie, Christmas berry
A much-branched shrublet up to 800 mm high with small, softish, linear leaves about 10 mm long and shiny, pink flowers (*ca.* 20 mm wide). Bright, orange-red berries cover the bushes during the summer and autumn months. It is widespread in sandy and rocky areas in the south-western Cape as well as in Kwazulu-Natal and Namaqualand. The plant was once used as a blood purifier in the treatment of piles. (Nov to Feb)

Chironia tetragona
An erect plant up to 600 mm high with thick leaves of varying width and up to 25 mm long, growing in pairs along four-angled stems. The sticky flowers (*ca.* 30 mm wide) are pink with conspicuous yellow stamens. This is a coastal species occurring from the Cape Peninsula to Port Elizabeth. (Oct to Jan)

Orphium frutescens
A robust shrublet growing to 800 mm with bright green leaves of varying width and up to about 35 mm long. The pink flowers (*ca.* 40 mm wide) are twisted in the bud stage before folding outwards and the five twisted stamens lie bunched together on one side of the stigma. It occurs in coastal, often marshy areas from Clanwilliam to the Cape Peninsula and eastwards to George. (Nov to Feb)

Sebaea aurea
An erect, annual plant (60-300 mm tall) that branches at the top. It has well-spaced pairs of opposite, oval leaves (*ca.* 15 mm long) and white or yellow four-petalled flowers (*ca.* 10 mm wide) clustered at the top of the plant. This plant grows in sandy flats and lower slopes from Clanwilliam to Port Elizabeth. (Oct to Dec)

Chironia tetragona

Chironia baccifera (fruit)

Chironia baccifera (flowers)

Orphium frutescens

Sebaea aurea

APOCYNACEAE

Carissa bispinosa
lemoenbessie/doring

A spiny shrub, or semi-climber, growing to 1-5 m tall. It has hard, shiny leaves and sturdy forked spines. The small, white flowers are sweet-scented and are borne in umbels, and the red berries are edible. It is found in coastal thicket from the Cape Peninsula to Port Elizabeth as well as in tropical Africa. (Sept to Dec)

ASCLEPIADACEAE

The family Asclepiadaceae comprises plants that have a milky sap and forked fruits containing seeds with tufts of silky hairs.

Cynanchum obtusifolium
bobbejaantou, klimop

A wiry climber with a poisonous milky sap. It has soft, ovate leaves and small green and white flowers. The elongated forked fruit pods contain numerous seeds with hairy tufts. It climbs over shrubs and trees in coastal thicket from the Cape Peninsula to Port Elizabeth and further to Kwazulu-Natal. (All year)

Astephanus triflorus
klimop

A slender vine with wiry, spirally twisted stems and pairs of leaves along the stem. The small flowers are cream or pink with maroon bracts. It occurs in coastal or inland thickets where it climbs on shrubs and is distributed between Clanwilliam and Bredasdorp as well as in Namaqualand. (April to Aug)

Microloma sagittatum
waxcreeper, kannetjies, bokhoringkies

A slender, twining herb with smallish, arrow-shaped leaves (*ca.* 25 mm long) with recurved margins. The small, pink flowers (*ca.* 7 mm long) are tubular with the petals barely opening and are clustered together in groups of three to nine. The forked fruits give this plant its popular name (= buck horns). As well as occurring in this area, it is found from Namaqualand to the Cape Peninsula. (June to Oct)

CONVOLVULACEAE

Falkia repens

A dwarf, often matted perennial plant with small, heart-shaped, stalked leaves (*ca.* 8 mm long) and dainty, funnel-shaped, pale pink flowers. It is usually found in marshy areas between the Cape Peninsula and Port Elizabeth. (Aug to Jan)

Cynanchum obtusifolium

Astephanus triflorus (on Diosma guthriei)

Microloma sagittatum

Carissa bispinosa

Falkia repens

BORAGINACEAE

Lobostemon lucidus   pyjamabos/bush
A low, compact shrub, woody at the base, growing to about 200 mm. It has light, grey-green, stiff leaves (50-70 mm long, 5-8 mm wide) that have a rough touch, and salmon-pink, tubular flowers (15-22 mm long) occasionally shading to blue at the tips. Found in Elim fynbos and limestone areas, this species occurs between Bredasdorp and Riversdale. (Aug to Oct)

Lobostemon sanguineus
An eye-catching, stout shrub growing to 1,5 m tall with leaves 20-25 mm long and 15-20 mm wide. The tubular flowers are a striking crimson. This plant is endemic to the Bredasdorp area where it occurs in fynbos on acid sands. (Jan to May)

Lobostemon curvifolius
A sprawling to upright shrub about 600 mm high with silver-grey, hairy leaves (40-50 mm long, 10 mm wide). The funnel-shaped flowers are light blue to pink (20-25 mm wide) and are produced in large numbers. It is locally common in a variety of habitats including renosterveld and limestone flats between Caledon and Riversdale. (Sept to Oct)

VERBENACEAE

Plexipus cernuus
A much-branched shrublet up to 600 mm high with small, oval leaves that are toothed at the tips. The white flowers have slender, curved tubes that are tinged with mauve. Often found on limestone soils, it occurs from the Cape Peninsula to Port Elizabeth and northward to Kwazulu-Natal. (Mar to Sept)

Lobostemon lucidus

Lobostemon sanguineus

Lobostemon curvifolius

Plexipus cernuus

STILBACEAE

Stilbe ericoides
A straggling to erect shrub growing up to 800 mm with whorls of linear, ericoid leaves (*ca*. 8 mm long) that have inrolled margins. The small, pink to mauve flowers (*ca*. 5 mm long) are densely packed into a spike at the end of branches. It is found on sandy flats and limestone hills from the Cape Peninsula to Port Elizabeth. (April to Sept)

Xeroplana zeyheri
A rounded shrublet about 500 mm high with small, mauve flowers crowded into a terminal flowerhead. It is endemic to this area where it grows in fynbos on the lower mountain slopes. (July to Oct)

SOLANACEAE

Lycium cinereum   doringbos, Cape boxthorn, slangbessiebos
A spiny, densely branched and woody shrub up to 1,5 m tall. It has tufts of small, narrow and slightly succulent leaves, tubular mauve flowers and red to black fruit. The branches end in rigid thorns, and were once used for stock-proof hedges. Fence poles were cut from larger plants. It grows in coastal thicket in the area as well as in the Little Karoo, Eastern Cape and Kwazulu-Natal. (April to Oct)

Solanum quadrangularis   dronkbessie, dronktou
A prostrate shrub with long, soft stems which is often found scrambling over other shrubs in dune thicket. It has clusters of hanging, blue-purple flowers that are star-shaped with yellow anthers in the centre forming a cone. It occurs on coastal dunes from the Cape Peninsula to Port Elizabeth. (Dec to Mar)

Stilbe ericoides

Xeroplana zeyheri

Lycium cinereum

Solanum quadrangularis

LAMIACEAE

Leonotis leonuris wilde/klip/rooidagga
A shrub growing to 2 m high with lance-shaped leaves (up to 120 mm long). The bright orange tubular flowers (*ca.* 50 mm long) are velvety and have a long, arching, hooded upper lip. Flowers are clustered in whorls at intervals along the stem. The seeds of this plant have been used to treat headaches and bronchitis. (Mar to May)

Stachys aethiopica katpisbossie
A straggling, herbaceous and aromatic plant about 500 mm tall. It has toothed, hairy leaves (*ca.* 20 mm long, 8-12 mm wide) on a four-angled stem, and white to pink hooded flowers (*ca.* 20 mm long) with purple flecks on the large lower lip. A widespread species that occurs in fynbos and forest areas between the Cederberg and Port Elizabeth and further to tropical Africa. (Aug to Sept)

Salvia africana-lutea bruinsalie, strandsalie
A much-branched aromatic shrub growing to 2 m tall with densely-packed grey, finely hairy leaves. The orange-brown flowers are clustered at the ends of the stems and have a long, hooded upper lip. Plants occur in dune fynbos and other dry fynbos areas between the Gifberg and Port Elizabeth. (June to Dec)

RETZIACEAE

Retzia capensis heuningblom
An erect, stiff-looking shrub growing up to 2 m tall. It is densely covered with narrow leaves that have inrolled margins, and tips that are often flushed with orange. The red, tubular flowers (*ca.* 25 mm long) end in a dark purple ring with a white tip, resembling the ash of a burning cigar. It is the only species in this family. Distribution is restricted to mountain slopes in the area as well as in the Klein River and Hottentots Holland mountains. (Sept to Mar)

Leonotis leonurus

Salvia africana-lutea

Stachys aethiopica

Retzia capensis

SCROPHULARIACEAE

Harveya sp.　　　　　　　　　　　　　　　　　　　　rooi inkblom
A root parasite without leaves. It has clusters of tubular, slightly curved yellow and bright pink flowers that turn black on drying. This species is intermediated between *H. capensis* and *H. purpurea*, and was found in Elim fynbos in the Vogelvlei area. (Sept to Dec)

Hyobanche sanguinea　　　　　　　　　　wolwekos, inkplant, katnaels
A root parasite without leaves reaching about 150 mm in height. The clustered pink to red tubular, furry flowers have projecting white anthers forming the characteristic "katnaels" (cat's nails). It is found on flats and lower slopes from the Gifberg to Port Elizabeth. (June to Nov)

Sutera hispida
A spreading shrublet up to 400 mm with irregularly-toothed leaves about 10 mm long. The small, white to mauve flowers have a distinct, tubular calyx and a corolla comprising a narrow tube that opens into flaring lobes from which one or both of the stamens protrude. Flowers are borne at the end of the branches. It occurs from Clanwilliam to Port Elizabeth. (All year)

Sutera revoluta
An erect shrublet up to 450 mm high with ericoid leaves (8-10 mm long). The mauve flowers have a distinct, tubular calyx and a corolla comprising a narrow tube that opens into flaring lobes from which one or both of the stamens protrude. This species grows on mid to upper slopes from Clanwilliam to Uniondale and is also found in Namaqualand. (May to Dec)

Sutera revoluta

Sutera hispida

Harveya sp.

Hyobanche sanguinea

Limosella grandiflora tongblaar, blouwaterblommetjie
A water-loving, tufted, perennial plant with clusters of oval leaves on long stalks and roots growing from the branches. It has small, white to lilac, five-lobed flowers. Occasionally found in pools and marshes from Riversdale to Port Elizabeth and also elsewhere in South Africa. (Aug to Mar)

Jamesbrittenia stellata
Members of this genus are similar to those of *Sutera* in that they both have cylindrical corolla tubes that open into star-shaped flowers. However, in *Jamesbrittenia* the stamens are hidden inside the corolla throat. This species grows to about 200 mm high and the tiny leaves (1-2 mm long) occur in small clusters along the stem. The flowers (10-15 mm long) are pink with velvety white-yellow markings and are borne on long, slender stalks (10 mm long). It occurs on limestone areas from the Cape Peninsula to Bredasdorp. (All year)

Jamesbrittenia albomarginata
A much-branched shrub about 600 mm tall with small clusters of tiny leaves (*ca.* 3 mm long). The flowers (10-15 mm long) are white with brown markings and have long, slender stalks (10 mm long). It grows on limestone hills and coastal dunes from the Cape Peninsula to Port Elizabeth. (All year)

Jamesbrittenia calciphila
These gnarled, woody, bonsai-like plants have small clusters of tiny leaves (1-2 mm long) and white to pink-violet flowers (5-8 mm long) on slender stalks. This species is usually restricted to limestone habitats and occurs from Bredasdorp to Uitenhage. (Aug to April)

Jamesbrittenia calciphila

Limosella grandiflora

Jamesbrittenia stellata   *Jamesbrittenia albomarginata*

Zaluzianskya villosa drumsticks
A hairy, branched annual up to 300 mm high with narrow leaves about 35 mm long. The white to mauve flowers have a yellow centre and five spreading lobes that are deeply notched at their tips. Flowers are crowded at the ends of the branches. It is found on sandy flats and lower slopes from Namaqualand to the southern Overberg. (June to Nov)

Nemesia barbata bloubekkie, fluweeltjie
An annual up to 300 mm high with coarsely-lobed leaves about 20 mm long. One to several flowers occur at the tops of erect branches and have an upper four-lobed, white segment and a lower bright, deep blue lip. It is found in sandy soils from Clanwilliam to Swellendam. (Aug to Sept)

Nemesia versicolor weeskindertjies
An annual about 500 mm high with lance-shaped, toothed leaves. The spurred flowers are a mixture of white, pink and blue and are clustered at the ends of the branches. It grows in sandy flats from Clanwilliam to Knysna and also in Namaqualand. (Aug to Nov)

Manulea tomentosa
A densely hairy, sprawling shrublet up to 800 mm tall with slightly toothed, narrow leaves. The tubular, orange to brick-red flowers end in five finger-like lobes and are clustered towards the ends of erect stems. It occurs on sand dunes and lower slopes from the Cape Peninsula to Bredasdorp. (June to Dec)

Lyperia lychnidea (= *Sutera lychnidia*)
A soft, spreading shrublet up to 800 mm high with furry young branches and narrow leaves that have small teeth towards their tips. The green to yellow flowers have a long tube that ends in five finger-like lobes and are clustered near the ends of branches. A plant of sandy flats from Saldanha Bay to the southern Overberg. (Sept to Dec)

Zaluzianskya villosa

Lyperia lychnidia

Manulea tomentosa

Nemesia barbata

Nemesia versicolor

Hebenstretia cordata
A much-branched shrublet about 300 mm high with small (*ca.* 7 mm long) heart-shaped, slightly upturned leaves crowded together. The four-lobed, white flowers have an orange centre and the corolla tube is split down the front. Flowers are clustered toward the end of branches. Growing as a pioneer on coastal dunes, it is widely distributed from Port Elizabeth to Clanwilliam and further to Namaqualand and southern Namibia. (Nov to May)

Dischisma ciliatum
An upright or spreading annual or perennial plant growing up to 400 mm with narrow, toothed leaves about 10 mm long. The white flowers (*ca.* 12 mm long) have a corolla that is split down the front and are clustered at the ends of branches. It occurs on flats and slopes from Nieuwoudtville to Port Elizabeth. (Aug to Dec)

SELAGINACEAE

Selago serrata
A stout, perennial plant up to 900 mm tall with oval leaves (*ca.* 15 mm long) that have serrated margins. The blue, tubular flowers have five lobes and are densely crowded into heads at the ends of branches. It occurs on mountain slopes between Clanwilliam and Knysna. (Oct to Feb)

Selago aspera
A much-branched perennial plant up to 300 mm high with white, furry branches and leaves. Linear leaves (2-6 mm long) are crowded onto the stems and the white to mauve, tubular flowers (*ca.* 5 mm long) occur in dense spikes up to 50 mm long. It occurs on lower sandstone and limestone slopes from Bredasdorp to Port Alfred. (Oct)

PLANTAGINACEAE

Plantago crassifolia
A low plant up to 400 mm high with a woody stem and fleshy, leathery leaves about 30 mm long. The small, white flowers are densely arranged into spikes at the ends of stoutish stalks. It grows in coastal dunes from Saldanha Bay to Port Elizabeth and also in tropical Africa. (Nov to Feb)

DIPSACEAE

Scabiosa columbaria — wild scabious
A herbaceous, perennial plant up to 800 mm with tufts of variably lobed leaves (*ca.* 80 mm long). The white or mauve flowerheads (*ca.* 50 mm wide) are borne singly at the ends of a long stem and comprise individual florets that open in a regular sequence from the outside inwards. It occurs on sandy flats and slopes and is widespread in the Western and Eastern Cape to the uplands of tropical Africa, as well as in Europe and Asia. (Aug to Feb)

Hebenstreitia cordata

Scabiosa columbaria

Selago aspera

Dischisma ciliatum

Plantago crassifolia

Selago serrata

CAMPANULACEAE

❏ Roella rhodantha
A sprawling shrublet up to 200 mm high with spreading branches. The rigid leaves are fringed with hairs and often are toothed toward the tips. The striking pink to red flowers (*ca.* 3 cm long) have blue spots on the lobes and they occur singly. This species is endemic to the rocky mountain slopes of the Potberg. (Nov to Jan)

Roella arenaria
A sprawling shrub growing to 400 mm with toothed leaves (*ca.* 8 mm long) densely arranged right up the stem to the base of the terminal, solitary flower. The bell-shaped flowers (*ca.* 25 mm wide) are white to pale blue with darker markings in the throat. It grows in coastal sands in the Bredasdorp area, as well westwards to Malmesbury. (Dec)

Roella incurva
An erect shrub up to 400 mm high with hairy stems. The leaves are bunched and often have linear teeth at the tips. The white to blue-grey, bell-shaped flowers (*ca.* 2,5 cm long) have narrow lobes that are often marked with dark spots and occur singly in groups of two to three. It occurs on lower mountain slopes in the area, as well as in the Cape Peninsula and from Tulbagh to Swellendam. (Oct to Jan)

❏ Rhigiophyllum squarrosum
A rigid and densely-leafed, erect shrublet growing to about 500 mm high. The oval leaves (*ca.* 7 mm long) have pointed tips and the deep blue flowers (15-20 mm long) have a narrow tube. This rather uncommon plant occurs on south-facing sandstone slopes and is endemic to the Caledon and Bredasdorp areas. (Nov to Jan)

Roella rhodantha

Roella arenaria

Roella incurva

Rhigiophyllum squarrosum

Lightfootia rigida
A rigid, erect shrub up to 600 mm high with small, reflexed leaves. The small, whitish flowers have evenly arranged narrow and free petals that are splayed outwards. It occurs on sandy flats and dry, rocky slopes in the area and elsewhere from Ceres to Uitenhage. (Nov to April)

Prismatocarpus brevilobus
An erect or sprawling slender shrublet growing to 300 mm with linear, concave leaves (up to 30 mm long). The white to blue flowers (*ca.* 10 mm long) are funnel-shaped with lobes that are shorter than the tube. It grows on mountain slopes in the area and elsewhere between Clanwilliam and Ladismith. (Dec to May)

Prismatocarpus spinosus
A stout, erect shrub up to 1 m tall covered with white hairs. The oval leaves (*ca.* 10 mm long, 5 mm wide) have sharp tips, and the bell-shaped flowers are white to lilac. This plant is endemic to the southern Overberg where it grows on rocky summits. (Dec to Feb)

Wahlenbergia capensis
A slender, erect and hairy annual plant ranging from 150-800 mm in height. It has coarsely toothed leaves (*ca.* 30 mm long) and turquoise-blue, bell-shaped flowers that are borne on long, slender stems. It occurs on sandy slopes between Clanwilliam and Bredasdorp. (Sept to Dec)

Wahlenbergia calcarea
A small plant up to 300 mm high. The pale blue flowers are borne at the end of long stalks (*ca.* 250 mm long). It is endemic to limestone areas of the southern Overberg. (Sept to Nov)

Lightfootia rigida

Prismatocarpus brevilobus

Prismatocarpus spinosus

Wahlenbergia capensis

Wahlenbergia calcarea

LOBELIACEAE

Plants in this family have two-lipped flowers, with the upper lip of two small petals split to the base at the back, and the lower lip with three larger petals.

❑ Lobelia valida galjoenblom

This erect, herbaceous and striking plant grows to 600 mm. It has softish, coarsely-toothed leaves scattered along the branches, and deep blue flowers (*ca.* 25 mm long) that are crowded at the tops of branches forming an attractive display. This species, which is only conspicuous after fire, is restricted to coastal dune and limestone hills in the Bredasdorp to Riversdale areas. Locals maintain that flowering time coincides with the galjoen fishing season. (Nov to April)

Lobelia setacea wilde/wild lobelia

An erect to sprawling, herbaceous perennial with slender stems and narrow, linear leaves (*ca.* 10 mm long) with deep blue to violet flowers (*ca.* 15 mm long). Found on sandy flats and slopes, it occurs between the Cape Peninsula and Caledon. (Nov to April)

Lobelia pubescens

A soft, sprawling, hairy plant with fine, angular branches growing to 350 mm long. Both the oval, toothed leaves (10-20 mm long, 8-20 mm wide) and the white flowers (*ca.* 14 mm long) are borne on fine (15 mm) stalks. Stems, leaves and flowers are hairy. It grows between Tulbagh and Humansdorp. (All year)

Lobelia tomentosa

This erect, hairy plant grows up to 400 mm high and has basal sessile leaves (12-30 mm long, 1-4 mm wide) with toothed margins. Clusters of purple to violet flowers (16-20 mm long) are borne at the ends of tall, thin stalks (100-200 mm long). Stems, leaves and flowers are hairy. This species occurs from the Hottentots Holland mountains to Grahamstown and eastwards to Kwazulu-Natal. (All year)

Cyphia volubilis klimop, bergbaroe

A herbaceous perennial with slender stems that twine round low bushes. It has narrowish leaves (*ca.* 20 mm long) and the white to pink or pale mauve, star-shaped flowers have two lips of two and three petals respectively. A widespread species that occurs in many habitats, it is distributed between Clanwilliam and Riversdale, as well as in Namaqualand. (Aug to Oct)

Lobelia valida

Lobelia setacea

Lobelia pubescens

Cyphia volubilis

Lobelia tomentosa

ASTERACEAE

Anaxeton virgatum   poeierkwassie
A sparse, upright shrublet about 200 mm high. The leaves have rolled margins and are rough on the upper surface and woolly below. The small flowerheads (*ca.* 30 mm across) comprise dense clusters of white or brownish, tubular florets at the top of a slender, leafless stem. It is common on mountain slopes in the Caledon and Bredasdorp areas. (July to Oct)

Arctotheca populifolia   seepampoen, tonteldoek(blom)
A soft, sprawling, perennial up to 150 mm high with white, woolly leaves (*ca.* 60 mm long). Yellow flowers are borne on long, stoutish, curving stems and are cupped in several rows of green involucral bracts of which the innermost ring has membranous margins. This plant is a common pioneer on shifting and ephemeral coastal dunes from the Cape Peninsula to southern Mozambique. The felt on the leaves was used as tinder. (All year)

Arctotis acaulis   botterblom, gousblom
A perennial, hairy plant up to 250 mm tall with a tuberous root and virtually no stem. It has a basal tuft of softish, stalked leaves and bright yellow or orange flowers that are maroon on the underside. Directly underneath the flowerhead are several rows of involucral bracts with the outer ones curving backwards and the inner row having rounded, transparent, purplish tips. It grows on sandy flats and slopes between Nieuwoudtville and Swellendam, as well as in Namaqualand. (July to Sept)

Athanasia dentata
An upright, rigid shrub up to 900 mm high. The multiple, yellow flowerheads are flat topped and the involucral bracts are horny. It occurs in renosterveld and on coastal flats from the Cape Peninsula to Port Elizabeth. (Aug to April)

Athanasia trifurcata   kouterbossie
A shrub growing between 500 mm and 1 m high. The leaves have grey hairs on the underside and it has yellow, flat-topped, multiple flowerheads. This species occurs throughout the southwestern and southern Cape. (Late summer)

Arctotis acaulis

Anaxeton virgatum

Arctotheca populifolia

Athanasia trifurcata

Athanasia dentata

Berkheya barbata
A rigid, prickly shrub up to 600 mm tall. The spiny, leathery leaves (60 mm long, 30 mm wide) are shiny above and woolly below. The yellow flowerheads are surrounded by long, spiny bracts. It occurs in both coastal sandy flats and rocky mountainous areas from the Gifberg to Mossel Bay. (Aug to Jan)

Berkheya coriacea   disseldoring
A perennial, scrambling shrublet up to 600 mm high with spiny, leathery leaves that are woolly beneath. It has large, striking, yellow flowers. This plant is endemic to limestone areas between Bredasdorp and Mossel Bay. (Sept)

Chrysanthemoides monilifera   bietou, boetabessie
A dense, upright shrub often growing taller than 2 m, with dark green, leathery leaves that are often toothed. It is frequently covered with yellow flowers (*ca.* 20 mm across) forming splashes of colour on coastal dunes and along roadsides. The purple berries are dispersed by birds. This widespread plant has numerous varieties and occurs in many areas of southern and sub-tropical Africa. The ash of this plant was used in soap manufacture. (All year)

Chrysanthemoides incana   strandbietou
A low, straggling shrub with branch tips that may be spiny. The leathery leaves tend to be woolly and greyish. The flowers are yellow. It occurs on coastal dunes and sandy inland slopes from Clanwilliam to Bredasdorp, as well as in Namaqualand and Namibia. (All year)

Chrysocoma coma-aurea
A woody shrublet up to 500 mm high with sparse, thin stems and narrow, ericoid leaves. The solitary, yellow flowerheads that occur at the ends of branches comprise disc florets that are all alike - there are no petal-like ray florets. There are several rows of green involucral bracts. It occurs on flats and slopes between Paarl and Caledon, and in this area it grows in habitats ranging from mountain slopes to coastal dunes. (Oct to Jan)

Berkheya barbata

Berkheya coriacea

Chrysanthemoides monilifera

Chrysanthemoides incana

Chrysocoma coma-aurea

Castalis nudicaulis
duinegousblom, witmagriet

A perennial herb up to 400 mm with toothed leaves (*ca.* 100 mm long, 15 mm wide) clustered at the base of the stem. The dark-centred white flowers have a maroon reverse and are borne at the end of a leafless stem. It occurs on mountain slopes and sandy flats from Clanwilliam to Uniondale. (Aug to Sept)

Cineraria geifolia
A scrambling plant up to 600 mm tall with round, irregularly-toothed leaves. The small (*ca.* 10 mm wide), bright yellow flowers are clustered towards the end of branches and have an unpleasant scent. It is common on coastal dunes from the Cape Peninsula to Kwazulu-Natal. (Aug to Nov)

Corymbium glabrum var. glabrum
This species grows to 400 mm and has tufts of stiff, broadly-linear (*ca.* 7 mm wide) leaves. The deep purple flowerheads are seen in abundance after fire when it rapidly regrows from a short underground stem. It grows on acid sands and clays from the Cederberg to Grahamstown. (Nov to Jan)

Corymbium africanum subsp. scabridum var. gramineum
heuningbossie

A tufted plant up to 300 mm high with a short underground stem clothed with long, soft hairs, and very narrow, stiff leaves (<1.5 mm wide). The white to pink-mauve tubular florets are grouped into flat-topped inflorescences. Flowering is common after fires. It occurs on flats and slopes from the Cederberg to Bredasdorp and also on the Swartberg mountains. (Oct to Jan)

Cotula filifolia
A slender annual plant up to 200 mm tall with wavy, linear leaves and a single white or yellow disc-like flowerhead. It is common in ditches and swampy areas in the area as well as westwards to Darling. (Aug to Nov)

Cullumia squarrosa
A robust, bristly shrublet about 350 mm tall (occasionally taller) with spiny, curved leaves covering the branches right up to the terminal, solitary flowerheads. The yellow flowers are cushioned on rings of spine-tipped involucral bracts. It occurs in dune fynbos between the southern Overberg and the Cape Peninsula. (June to Oct)

Cullumia squarrosa   *Castalis nudicaulis*

Corymbium glabrum var. *glabrum*

Cotula filifolia

Corymbium africanum subsp. *scabridum* var. *gramineum*

Cineraria geifolia

Edmondia sesamoides (wit)sewejaartjie
A sparsely branched shrublet up to 600 mm tall with short leaves (5-8 mm long, 1 mm wide) held flat against the upper parts of the stem and longer leaves (20-40 mm) lower down. The flowers are pink in bud opening to white, occasionally yellow. It is common on sandy and rocky flats from the Cape Peninsula to Riversdale. (Aug to Jan)

❑ **Dymondia margaretae** vlei daisy
A perennial plant forming mats about 50 mm high. The leaves (25-30 mm long,. 3 mm wide) are waxy green above with a dense, white felt underneath. The small, yellow flowers are 25 mm wide. It is endemic to the southern Overberg where it grows on sandy coastal flats as well as in seasonally wet pans and vleis. It withstands long periods of inundation. This charming plant has become popular as a ground cover in gardens. (All year)

Stoebe plumosa slangbos
A much-branched shrub up to 1,2 m high with clusters of tiny, woolly leaves pressed to the stems giving the plant an overall silver-grey, granular appearance. The purple to brown tubular florets are arranged in a long spike at the ends of the branches. This widespread plant occurs on coastal dunes and other habitats between the Gifberg and Port Elizabeth. (April to June)

Dimorphotheca pluvialis Cape daisy, witmagriet
An annual up to 400 mm high. The lance-shaped leaves (*ca.* 70 mm long) are bluntly toothed. The flower heads (*ca.* 60 mm wide) have white petal-like ray florets that are violet on the underside, and there is a single row of separated, narrow involucral bracts. This daisy forms brilliant white sheets in disturbed, sandy areas in the spring and is common in this area, as well as between the Cape Peninsula and Namaqualand. (Aug to Oct)

Disparago anomala
A much-branched shrublet up to 200 mm with small, narrow, ericoid leaves that are often twisted. The tiny, white or pink florets are crowded into flowerheads at the ends of the branches. If examined closely two kinds of florets can be teased out - those with a single strap-shaped petal, and those without. It is fairly common in coastal, sandy habitats between the southern Overberg and the Cape Peninsula. (Dec to April)

Stoebe plumosa *Disparago anomala* *Edmondia sesamoides*

Dymondia margaretae

Dimorphotheca pluvialis

Eriocephalus paniculatus kapokbossie, wild rosemary
An upright shrub about 1 m tall with fairly lax, thin (1-2 mm across) branches and clusters of small, silvery, linear leaves (8-17 mm long, 0,4-0,8 mm wide) that are sometimes divided into one to three segments. The small flowerheads are white to pale purple and become woolly with age. This widespread species occurs in fynbos, renosterveld and karoo habitats from Vanrhynsdorp to Humansdorp and also further to the interior. (July to Sept)

Euryops abrotanifolius geelmagriet
A robust, erect shrub up to 1 m with slightly fleshy leaves that are deeply divided into narrow, linear segments. Typical of this genus, the bright yellow, conspicuous flowerheads (*ca.* 50 mm wide) occur singly on naked stalks and there is a single row of involucral bracts that are united at the base forming a flat-bottomed cup. It grows on mountain slopes between Clanwilliam and Swellendam. (May to Dec)

❏ **Euryops linearis**
A large, erect shrub growing to about 2 m high with straight leaves (30-40 mm long, 1-2 mm wide) and yellow flowerheads (10-12 mm wide). This plant is endemic to limestone areas between Bredasdorp and De Hoop. (Aug to Oct)

Elytropappus rhinocerotis renosterbos(sie)
A much-branched shrub growing to about 1 m tall. It has minute, ericoid leaves and tiny, inconspicuous yellowish flowers. This plant is easily recognised by its drab, grey-green colour in renosterveld and degraded Elim fynbos. Its colour is believed to be the reason for its common name (renosterbos = rhinoceros bush). In the "ruensveld" north-west of Bredasdorp, where there are few trees, this plant was widely used as fuel for stoves. It was also used medicinally during the 1918 influenza epidemic. (Mar to Sept)

Eriocephalus paniculatus
(seed and flowers)

Elytropappus rhinocerotis

Euryops linearis

Euryops abrotanifolius

Tarchonanthus camphoratus kapokbossie, kanferboom
An aromatic shrub or tree growing up to 3 m high, with grey-green, oval leaves that have a whitish reverse. The small, creamy flowerheads are borne in clusters that later turn to cottonwool-like bunches of seed that are used by birds to line their nests. It is widespread plant that grows in a wide range of habitats throughout much of sub-Saharan Africa. In the southern Overberg this tree is most common in coastal thicket. (Dec to April)

Gazania pectinata (ogies)gousblom
An annual or perennial up to 400 mm. The basal leaves are highly divided and are green on the upper surface with white, woolly undersides. The large, showy flowerheads (*ca.* 90 mm wide) are yellow to orange, occasionally with a dark ring or centre and there are several rows of involucral bracts that are united at the base to form a cup. It is found in a variety of habitats in the area and westwards to the Cape Peninsula. (Aug to Nov)

Felicia amoena subsp. latifolia
A biennial or perennial plant up to 250 mm high. The leaves (50-70 mm long; 8-12 mm wide) are hairy and the deep blue flowers (*ca.* 30 mm wide) have yellow centres. It grows in dune and limestone fynbos in the area and also occurs elsewhere between the Cederberg and Port Elizabeth. (June to Nov)

Felicia aculeata
A perennial shrub up to 450 mm tall with hairy leaves (5-12 mm long, 2-3 mm wide) and blue flowers (*ca.* 20 mm wide) with yellow centres. It occurs on acid and limestone-derived sands in the area and is widely distributed between Tulbagh and Knysna. (Aug to Sept)

Oedera capensis
A sparse shrub up to 300 mm tall with crowded, stiff, narrow, spiny leaves (*ca.* 20 mm long). The compound flowerhead consists of many. small flowerheads crowded together into the larger bright yellow "flower". Beneath are several rows of rather translucent involucral bracts. It grows on dry flats and slopes between the Cape Peninsula and Riversdale, as well as in the Little Karoo. (June to Oct)

Oedera genistifolia
A compact, low shrub about 500 mm high with clusters of small, yellow compound flowerheads. It grows in clayey renosterveld areas and also on limestone hills in the area. (August to Dec)

Gazania pectinata

Tarchonanthus camphoratus

Oedera genistifolia   *Oedera capensis*   *Felicia aculeata*

Felicia amoena subsp. *latifolia*

Helichrysum crispum (hottentots)kooigoed
A sprawling perennial herb up to 600 mm tall with soft, white, silky-woolly leaves (*ca.* 35 mm long, 12 mm wide) that clasp the stem. The creamy-white flowerheads (*ca.* 5 mm wide) are densely clustered at the ends of branches and have distinctly crisped involucral bracts. It grows in coastal dunes in the area as well as westward to the Cape Peninsula. Concoctions made from leaves have been used as a cure for high blood pressure. (Oct to Jan)

Helichrysum dasyanthum
An erect, spreading plant forming large, lax bushes up to 1 m high with softly hairy leaves. There are terminal clusters of yellow flowerheads. It grows on coastal and mountain slopes between Clanwilliam and Uniondale as well as in the Little Karoo. (Sept to Dec)

Helichrysum retortum
An attractive shrublet spreading along the ground with small, silvery leaves and shiny white flowers that have maroon tips. It is common in coastal dunes and rocky shores between the southern Overberg and the Cape Peninsula. (Aug to Jan)

Helichrysum teretifolium
A soft, erect, well-branched shrub growing to 400 mm tall. It has dark green, ericoid leaves and terminal clusters of small, cream or rosy flowers with yellow centres. Plants grow on dunes and coastal slopes between the Cape Peninsula and southern Kwazulu-Natal. (July to Dec)

Helichrysum chlorochrysum
An erect shrub up to 1 m tall with silvery leaves and cream to yellow flower clusters at the tips of branches. This species is endemic to limestone fynbos between Bredasdorp and Riversdale.

Helichrysum crispum

Helichrysum teretifolium

Helichrysum dasyanthum

Helichrysum retortum

Helichrysum chlorochrysum

Syncarpha canescens (= *Helipterum canescens*) rooisewejaartjie
An erect to straggling shrublet up to 300 mm high with small, silver leaves closely pressed to the stem. The everlasting flowers are brilliant pink to red. It occurs on sandy flats, slopes and limestone areas between Gifberg and Humansdorp. These plants were once extensively exported for filling mattresses and for making wreaths. Other species of *Helichrysum, Edmondia* and *Syncarpha* were also used for these purposes. (Jan to Sept)

Syncarpha vestita (= *Helichrysum vestitum*) strooiblommetjie, matras sewejaartjie, Cape everlasting
An erect, robust soft shrub up to 1 m, with soft, furry leaves. The solitary, showy, white flowerheads (*ca.* 40 mm wide) have maroon and white centres, and are borne at the end of woolly stems. Plants are common in the first few years after fire and are found on flats and slopes between the Cape Peninsula and Knysna and also in the Swartberg mountains. (Nov to Jan)

Syncarpha paniculata (= *Helichrysum paniculatum*) sewejaartjie
Upright shrubs about 1 m tall with narrow silver-grey leaves. The flowerheads are pink in bud and open to white or pale yellow with yellow centres. It occurs on flats and lower slopes from Bredasdorp to Port Elizabeth. (All year)

Syncarpha argyropsis (= *Helipterum argyropsis*) witsewejaartjie, beesoogsewejaartjie
Compact, rounded bushes up to 700 mm tall with soft silver-grey leaves. The white everlasting flowers have maroon markings and yellow centres. It grows on rocky coastal hills between Bredasdorp and Riversdale. (Sept to Dec)

Syncarpha canescens

Syncarpha vestita

Syncarpha paniculata

Syncarpha argyropsis

The genus *Metalasia* comprises shrubs that usually have woolly stems and have linear, ericoid leaves that are often spirally twisted or tufted. The flat-topped compound flowerheads are multiple units of smaller flowerheads which in turn comprise bundles of three to ten or more florets.

Metalasia brevifolia blombossie
A highly-branched plant up to 1,2 m tall with bunches of dwarfed leaves and white flowerheads. It grows on sandy flats and slopes between Nieuwoudtville and Port Elizabeth. (Oct to Feb)

Metalasia calcicola
This species, which is restricted to limestone of the southern Cape is about 400 mm high. The stems are densely packed with tufts of small, curved, dark green leaves. The flowerheads are also densely packed with white florets. (Oct to May)

Metalasia muricata blombos, steekbos
Sturdy plants growing up to about 1-2 m, with hard, sharpish, twisted leaves. This winter-flowering plant is common in the area, brightening up the road verges with its white flowerheads. It is widely distributed in fynbos areas from Clanwilliam to Port Elizabeth and also further to the interior. (April to Sept)

Metalasia pungens blombos
An erect, rigid plant growing to 2,5 m high with branches that are woolly when young and later become smooth. The spreading, non-twisted leaves may be straight or upward-curving and the dense flowerheads are white to yellowish-brown. It grows in a variety of fynbos habitats, including limestone and also in renosterveld. (March to Aug)

Metalasia serrata rooiblombossie
An erect plant up to 600 mm high with branches that are densely woolly when young and later become smooth. The dense foliage comprises half-twisted leaves that are straight to slightly curved. The flowerheads are whitish to reddish-pink or mauve. This species is confined to sandstone slopes from Stanford to Bredasdorp. (July to Dec)

Metalasia pungens

Metalasia muricata

Metalasia serrata

Metalasia calcicola

Metalasia brevifolia

Osteospermum fruticosum
rankmagriet

A spreading, mat-like perennial with oval, succulent leaves. The white flowers have a lilac centre and are purple on the reverse. There are several rows of soft, free, involucral bracts. It is common at the coast, often growing close to the high water mark and occurs between Malmesbury and Kwazulu-Natal. (June to Oct)

Osteospermum subulatum

This leafy, prostrate perennial has long trailing stems and single terminal, yellow flowers, often with a maroon reverse. It is endemic to limestone hills and flats of the Bredasdorp region. (Aug to April)

Othonna filicaulis

A straggling, branched plant up to 700 mm tall with heart-shaped leaves that encircle the stem at their bases. The yellow, occasionally white or pink, flowerheads have no "petals" (ray florets) and are borne at the ends of protruding stalks. It occurs on sandy flats and slopes from Clanwilliam to Bredasdorp. (June to Aug)

Othonna dentata

A low, succulent shrub sometimes growing to 700 mm, but it is usually shorter when it occurs close to the high water line. The leaves (*ca.* 50 mm long) are often coarsely toothed. Yellow flowerheads occur singly at the end of naked stems, and there is one row of more-or-less united involucral bracts. It is common in coastal sandy areas and on inland mountain slopes between the Riversdale area and the Cape Peninsula. (May to Dec)

Othonna quinquedentata

An erect, sparsely branched shrub up to 1,5 m. The fleshy leaves (up to 150 mm long) are clustered towards the base of the plant, and are usually coarsely toothed. The yellow flowers (*ca.* 30 mm wide) are borne at the ends of long, leafless stems. It occurs in damp, acid sands from the Cape Peninsula to Mossel Bay as well as further inland to the Worcester area and the Swartberg mountains. (All year)

Othonna quinquedentata

Osteospermum subulatum

Osteospermum fruticosum

Othonna filicaulis

Othonna dentata

Phaenocoma prolifera rooisewejaartjie, Cape everlasting
A striking, densely branched, rigid shrub up to 600 mm high. The minute leaves occur in small clusters on short side branches, giving the plant its characteristic "granular" appearance. The showy flowerheads are a bright pink-red. Involucral bracts occur in many rows with the innermost row forming the pink papery everlasting "petals". It grows in acid sands between the Cape Peninsula and Bredasdorp and also inland to the Klein Swartberg mountains. (Sept to March)

Lachnospermum imbricatum
A shrub growing to 500 mm high with ericoid leaves and flowerheads of small, purple florets. It occurs on limestone flats and slopes between the Cape Peninsula and Bredasdorp. (Jan to Mar)

Polyarrhenia stricta
A prickly shrub up to 300 mm high with sharp-pointed, rough leaves (*ca.* 10 mm long). The flower heads occur singly at the ends of branches and have white "petals" with a purple reverse, and yellow centres. It grows in acid sands in the southern Overberg. (July to Sept)

Pteronia incana laventelbossie, perdebossie
A much-branched, shrub up to 1 m high, with small, narrow grey-green leaves (*ca.* 10 mm long). The light yellow flowerheads (*ca.* 6 mm wide) lack a fringe of "petals" and consist of tubular disc florets only. It occurs in renosterveld between Namaqualand and Port Elizabeth.

Ursinia paleacea geelmagriet
A herbaceous shrublet up to 900 mm high with finely-lobed leaves (*ca.* 50 mm long). The solitary, yellow flowerheads are borne at the ends of slender, bare stems and the ray "petals" have dark markings at the base. There are several rows of fairly broad involucral bracts with the innermost ones having conspicuous and characteristic papery tips. It occurs on damp mountain slopes between the Cape Peninsula and Humansdorp.

Ursinia paleacea

Pteronia incana

Polyarrhenia stricta

Lachnospermum imbricatum

Phaenocoma prolifera

Senecio elegans wild cineraria, strandblommetjie
An annual plant up to 500 mm tall, occasionally taller to 1 m. The lobed and divided leaves (*ca.* 80 mm long) are hairy, and the flower heads (25 mm wide) are purple with yellow centres and there is one row of involucral bracts. It occurs in dune fynbos and other sandy flats and slopes from Nieuwoudtville to Port Elizabeth and also in Namaqualand. (July to Mar)

Senecio sophioides
A soft, slightly hairy, herbaceous annual plant with divided leaves growing up to 300 mm. The flowers may be deep blue, purple or yellow. Plants grow in sandy flats and slopes from the Cederberg to Swellendam. (July to Oct)

Senecio burchellii geelgifbossie
A branching, woody shrublet up to 300 mm high. The narrow leaves have margins that are rolled under and the flowerheads are yellow. This plant is common in disturbed, sandy areas from Clanwilliam to Port Elizabeth. It owes its common name (= yellow poison bush) to the fact that it is poisonous to stock. (All year)

Senecio arenarius hongerblom
A branching, annual up to 400 mm high with deeply-lobed leaves (*ca.* 70 mm long) that have glandular hairs. The purple flowerheads (*ca.* 250 mm wide) have yellow centres and there is one row of involucral bracts, with a few dark-tipped bracts at the base. It occurs in sandy flats between southern Namibia and Bredasdorp. (Aug to Oct)

Senecio elegans

Senecio sophioides

Senecio burchellii

Senecio arenarius

Glossary

alternate: applied to leaves placed on opposite sides of the stem at different levels
annual: a plant completing its life cycle within one year
anther: upper portion of the stamen which produces the pollen (fig. 6)
axil: the upper angle between the stem and the leaf stalk (fig. 1)

Fig. 1   leaf arrangement

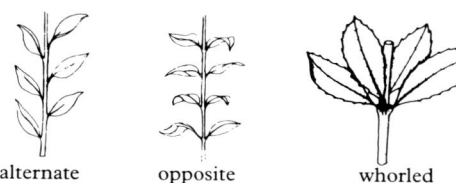

alternate   opposite   whorled

berry: fruit in which the seeds are enclosed in juicy pulp
bipinnate: when divisions of a pinnate leaf are themselves pinnate (fig. 2)
bract: a modified leaf subtending a flower or a flower stalk
bulb: an underground storage organ composed of fleshy leaf bases or scales
calyx: outer envelope of the flower consisting of sepals (fig. 6)
capsule: a dry fruit which splits open
cordate: heart-shaped in outline
corm: an underground storage organ which is a modified stem
corolla: inner envelope of the flower consisting of free or united petals (fig. 6)
corona: circle of appendages between corolla and stamens often forming a ring or crown
crenate: scalloped or notched
crisped: curled, as when the edge is excessively and irregularly divided and twisted

Fig. 2   leaf types

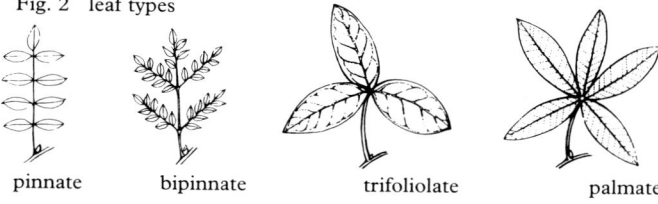

pinnate   bipinnate   trifoliolate   palmate

cyme: a flattened branched inflorescence in which the central flower opens first (fig. 5)
deciduous: seasonal total leafshed, usually in autumn or winter
dentate: toothed, notched (fig. 4)
disc-floret: tubular flower in the central part of the flowerhead in Asteraceae
distichous: arranged regularly in two opposite rows like a fan
entire: an even margin on a leaf, without teeth or notches (fig. 4)
epicalyx: a whorl of bracts round a flower, like an extra calyx
epiphyte: a plant which grows on another plant but not parasitically
evergreen: shrub or tree always in leaf

exserted: protruding
filament: stalk supporting the anther (fig. 6)
floret: tiny flower, one of a cluster
follicle: a dry fruit opening on one side only
glaucous: pale bluish–green or with a pale bloom

Fig. 3   leaf forms

linear   lanceolate   obovate   ovate

herb: a non–woody plant
inflorescence: the arrangement of the flowers on the floral axis
keel: the two lower petals of a pea–shaped flower folded like the keel of a boat
lanceolate: lance–like, longer than broad and tapering towards the apex (fig. 3)
linear: long and narrow with almost parallel margins (fig. 3)
lobe: part of a leaf, calyx or petal formed by a division to about the middle
midrib: central and most prominent vein in a leaf

Fig. 4   leaf margins

sentire   dentate   undulate

node: the point on a stem where a leaf is or may be borne
obovate: reversed ovate, the broadest part above the middle (fig. 3)
opposite: as of leaves when two are borne at the same node on opposite sides of the stem (fig 1)
ovary: that part of the flower that contains the ovules and eventually becomes the fruit
ovate: egg–shaped in outline, with the broadest part below the middle (fig. 3)
palmate: divided or lobed like a hand (fig. 2)
panicle: a branching raceme (fig. 5)
parasitic: growing on and getting food from a host plant
pedicel: stalk of each single flower on an inflorescence
peduncle: stalk bearing a cluster of flowers
peltate: as of a leaf attached by its lower surface to a stalk, instead of by its margin
perennial: a plant that lives for more than two years, usually flowering every year
perianth: floral envelopes consisting of calyx, corolla or both, or tepals
petal: one unit of the corolla, usually brightly coloured (fig. 6)
petiole: a leaf stalk

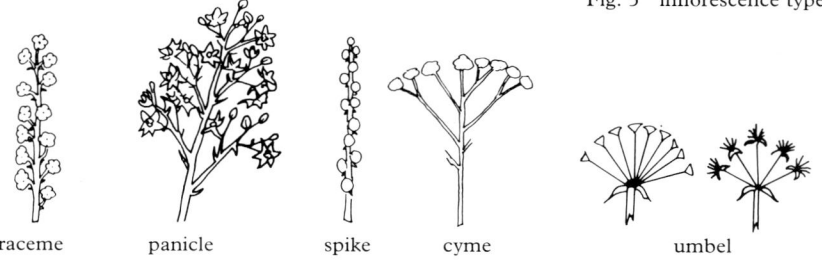

Fig. 5 inflorescence types

raceme   panicle   spike   cyme   umbel

pinnate: with the leaflets of a compound leaf arranged on each side of a common petiole (fig. 2)
procumbent: lying along the ground
raceme: an inflorescence in which the flowers, progressively younger upwards, are borne on pedicels along an elongated axis (fig. 5)
ray-florets: narrow, usually strap–shaped flowers on the margin of the flowerhead in Asteraceae
rhizome: a root–like stem, prostrate on or under the ground, producing roots and new shoots at the tip
scandent: a climber which leans on other plants
sepal: a segment of the calyx, usually green (fig. 6)
serrate: with teeth on the margin like a saw
sessile: without a stalk
sheath: tubular structure enveloping the base of a leaf or a bud
spadix: a large bract enclosing a flower cluster
spike: similar to a raceme but with sessile flowers (fig. 5)
spur: a slender, usually hollow, projection of a flower often carrying nectar
stamen: unit of flower composed of filament and anther (fig. 6)
staminode: an imperfect stamen, infertile and often looking like a petal
stigma: the part of the style which receives the pollen (fig. 6)
tepal: a segment of those perianths not clearly differentiated into typical calyx and corolla, as in the families Liliaceae and Iridaceae
terrestrial: on or in the ground
trifoliolate: a compound leaf with three leaflets (fig. 2)
tuber: thickened underground stem or root acting as a food reservoir
umbel: inflorescence in which the flower stalks spring from one point; compound umbel where each ray itself bears an umbel (fig. 5)
undulate: wavy as in margins (fig. 4)
whorl: arrangement of three or more leaves or flowers in a circle round an axis (fig. 1)

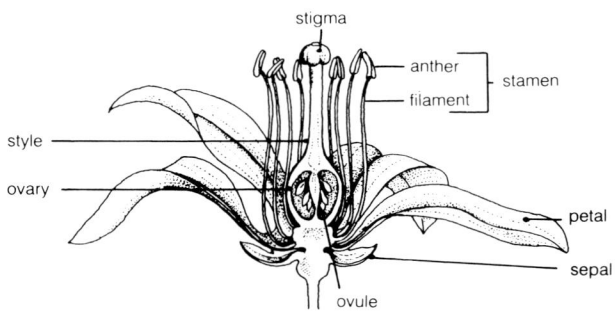

Index

aambeibossie 206
aandblom 86
aandgonna 180
aand kaneelbol 154
aandpypie 94
aardroos 118
aasbossie 160
Acacia cyclops A. Cunn. ex G. Don 142
Acacia longifolia (Andr.) Willd. 142
Acacia mearnsii De Wild. 142
Acacia saligna (Labill.) Wendl. 142
Acmadenia mundiana Eckl. & Zeyh. 160
Acmadenia obtusata (Thunb.) Bartl. & Wendl. 160
Adenandra gummifera Strid 160
Adenandra obtusata Sond. 160
Adenandra viscida Eckl. & Zeyh. 160
Adromischus caryophylllaceus (Burm. f.) Lem. 136
Agathosma 162
Agathosma cerefolium (Vent.) Bartl. & Wendl. 162
Agathosma collina Eckl. & Zeyh. 162
Agathosma dielsiana Schltr. ex Dummer 162
Agathosma riversdalensis Dummer 162
Agathosma serpyllacea Licht. ex Roem. & Schultes 162
Agulhas conebush 108
AIZOACEAE 124
Aizoon rigidum L.f. var **angustifolium** Sond. 124
Albuca maxima Burm. f. 60
Aloe arborescens Mill. 60
Aloe brevifolia Miller 60
altydbossie 138
AMARYLLIDACEAE 68
Amaryllis belladonna L. 68
Amphithalea alba 144

Amphithalea bicvulata (H. Bol.) Granby 144
Amphithalea ericifolia Eckl. & Zeyh. subsp **ericifolia** 144
ANACARDIACEAE 170
Anapalina nervosa 92
Anaxeton virgatum DC. 230
Anemone tenuifolia (L. f.) DC.132
Anginon difforme (L.) B.L. Burtt 184
Anisodontea scabrosa (L.) Bates176
Anomalanthus scoparius Klotzsch 186
Anomalesia cunonia 80
anysboegoe 162
APIACEAE 184
APOCYNACEAE 208
APONOGETONACEAE 38
Aponogeton angustifolius Ait. 38
Aponogeton distachyos L.f. 38
April fool 70
ARACEAE 46
ARALIACEAE 182
Arctopus echinatus L. 184
Arctotheca populifolia (Berg.) T. Norl. 230
Arctotis acaulis L. 230
Aristea glauca Klatt 74
Aristea oligocephala Bak. 74
arrow reed 48
arum lily 46
ASCLEPIADACEAE 208
Aspalathus calcarea Dahlg. 144
Aspalathus caledonensis Dahlg. 144
Aspalathus hirta E. May subsp **hirta** 146
Aspalathus incurvifolia J.R.T. Vogel ex Walp. 146
Aspalathus juniperina Thunb. 146
Aspalathus pycnantha Dahlg.146
Aspalathus securifolia Eckl. & Zeyh. 146
ASPARAGACEAE (Liliaceae) 64
Asparagus asparagoides (L.) Druce 64

257

ASPHODELACEAE (Liliaceae) 58
Astephanus triflorus (L. f.) Schultes 208
ASTERACEAE 230
Athanasia dentata (L.) L. 230
Athanasia trifurcata (L.) L. 230
Augustusbossie 166
Aulax umbellata (Thunb.) R. Br. 100
awl–leaf sugarbush 118

Babiana montana G.J. Lewis 74
Babiana patersoniae L. Bol. 74
Babiana patula N.E. Br. 74
Bartholina burmanniana (L.) Ker–Gawl. 96
Bassia diffusa (Thunb.) Kuntze 124
basbessie 122
bashful sugarbush 118
basterkreupelhout 114
bearded disa 96
beesbossie 148
beesoogsewejaartjie 244
belladonna 68
bergbaroe 228
bergkatjiepiering 104
berglelie 68
berg (mountain) bluebell 80
bergpalmiet 44
bergpypie 80
bergseldery 184
bergtee 152
Berkheya barbata (L. f.) Hutch. 232
Berkheya coriacea Harv 232
Berzelia abrotonoides (L.) Brongn. 138
Berzelia lanuginosa (l.) Brongn. 138
besembiesie 76
besemriet 52
biesie 76
bietou 232
black wattle 142
blisterbush 184
bloedblom 70
blombos 246

blombossie 246
blou afrikaner 80
bloubekkie 220
blouertjieboom 166
bloupypie 82
blouwaterblommetjie 218
blouwaterlelie 132
Bobartia longicyma J.B. Gillett subsp. **magna** J.B. Gillett ex Strid 76
bobbejaantjie 74
bobbejaantou 66, 208
boetabessie 232
bokhoringkies 208
Bonatea speciosa (L.f.) Willd. 98
Boophane disticha (L.f.) Herb. 68
BORAGINACEAE 210
bosluisgras 40
bottelheide 188, 200
bottelheide/heath 192
botterblom 230
botterpypie 86
bottle heath 188, 192, 200
Bot River protea/sugarbush 120
Brachysiphon acutus (Thunb.) Juss. 178
brakblommetjie 202
BRASSICACEAE 134
Bredasdorp conebush 104
Bredasdorp lily/lelie 70
Bredasdorp protea 120
Bredasdorp sceptre 110
broad–leaf featherbush 100
brown–bearded protea/sugarbush 118
bruinsalie 214
brunia 138
BRUNIACEAE 138
Brunia laevis Thunb. 138
Brunsvigia orientalis (L.) Ait. ex Eckl. 68
buffalo grass 42
buffelskweek 42
Bulbine lagopus (Thunb.) N.E.Br. 58
Bulbinella nutans (Thunb.) Dur. & Schinz var. **nutans** 58

buxifolium 116
Calopsis vimineus (Rottb.) Linder 48
CAMPANULACEAE 224
cancer bush 150
candelabra flower 68
Cape box–thorn 212
Cape daisy 236
Cape edelweiss 66
Cape everlasting 244, 250
Cape may 160
Cape myrtle 202
Cape smilax 64
Carissa bispinosa (L.) Desf. ex Brenan 208
Carpobrotus acinaciformis (L.) L. Bol. 126
Carpobrotus edulis subsp. **edulis** (L.) L. Bolus 126
CARYOPHYLLACEAE 132
Caryotophora skiatophytoides Leistner 126
Cassine maritima (H. Bol.) L. Bol. 172
Cassine peragua L. 172
Cassytha ciliolata Nees 134
Castalis nudicaulis (L.) T, Nori 234
CELASTRACEAE 172
Centella virgata (L.f.) Drude 184
Ceratocaryum argenteum Nees ex Kunth 48
Chasmanthe aethiopica (L.) N.E. Br. 76
Chenolea diffusa 124
CHENOPODIACEAE 124
cherrywood 172
china flower 160
chinkerinchee 64
Chironia baccifera L. 206
Chironia tetragona L. f. 206
Chondropetalum microcarpum (Kunth) Pillans 48
Chondropetalum tectorum (L.f.) Rafin 48
Christmas berry 206
Christmasblom 100

Chrysanthemoides incana (Burm. f.) T. Norl. 232
Chrysanthemoides monilifera (L.) T. Norl 232
Chrysocoma coma–aurea L. 232
Cineraria geifolia (L.) L. 234
Cissampelos capensis L. f. 134
Cliffortia feruginea L. f. 140
Cliffortia ilicifolia L. var **ilicifolia** 140
Cliffortia strobilifera L. 140
Cliffortia stricta Weim. 140
Clutia ericoides Thunb. 168
Coelidium ciliare (Eckl. & Zeyh.) Walp. 148
COLCHICACEAE (Liliaceae) 58
Coleonema album (Thunb.) Bartl. & Wendl.160
Colpoon compressum Berg. 122
Colpoon speciosum (A.W. Hill) Bean 122
common pagoda 110
common pin spiderhead 112
common stream conebush 108
common sugarbush 120
common sunshine conebush 106
compacta 120
Conicosia pugioniformis (L.) N.E. Br. subsp **muirii** (N.E. Br.) Ihlenfeldt & Gerbaulet 126
CONVOLVULACEAE 208
Corymbium africanum L. subsp **scabridum** (Berg.) Weitz var **gramineum** (Burm. f.) Weitz 234
Corymbium glabrum L. var **glabrum** 234
Cotula filifolia Thunb. 234
Cotyledon orbiculata L. var **orbiculata** 136
cowled friar 98
CRASSULACEAE 136
Crassula expansa Drand. subsp **filicaulis** (Haw.) Tolken 136
Crassula fallax Friedr. 136
Crassula nudicaulis L. var **nudicaulis** 136
Cullumia squarrosa (L.) R. Br. 234

Cussonia thyrsiflora Thunb. 182
Cyanella lutea L.f. 72
Cynodon dactylon (L.) Pers. 38
CYPERACEAE 44
Cyphia volubilis (Burm. f.) Willd. 228
Cynanchum obtusifolium L. f. 208
Cyrtanthus carneus Lindl. 70
Cyrtanthus guthrieae L.Bol. 70
Cyrtanthus leucanthus Schltr. 70

Dasispermum suffruticosum (Berg.) B.L. Burtt 184
dekriet 48, 56
Delosperma litorale (Kensit) L. Bol.128
Dianthus albens Ait.132
Dilatris pillansii W.F. Barker 66
Dimorphotheca pluvialis (L.) Moench 236
Diosma 164
Diosma guthriei P.e. Glover 164
Diosma haelkraalensis I. Williams 164
Diosma subulata Wendl. 164
DIPSACEAE 222
Disa cornuta (L.) Swartz 96
Dischisma ciliatum (Berg.) Choisy 222
Disparago anomala Schltr. ex Levyns 236
Disperis capensis (L.) Swartz 96
disseldoring 232
doringbos 212
Dorotheanthus bellidiformis (Burm.f.) N.E. Br. 128
doublom 134
Drimia media Jacq. 64
dronkbessie 212
dronktou 212
Drosanthemum hispidum (L.) Schwant. 128
Drosanthemum intermedium (L. Bol.) L Bol. 128
Drosera capensis L. 134
DROSERACEAE 134
Drosera cistiflora L. 134
drumsticks 220

duine-aalwyn 60
duinebuchu 160
duinegousblom 234
duineknoppiesbos 106
duinesuikerbos 120
dune bluebell 84
dune conebush 102
Dymondia margaretae Compton 236
early blue disa 96

EBENACEAE 204
Edmondia sesamoides (L.) Hilliard 236
Ehrharta villosa Schult. f..var. maxima Stapf 38
Elegia filacea Mast. 50
Elegia muirii Pillans 50
Elegia persistens Mast. 50
Elim conebush 104
Elim heath 198
Elimsheide 198
Elytropappus rhinocerotis (L. f.) Less. 238
Empodium gloriosum (Nel) B.L. Burtt 72
Erica 186, 188
ERICACEAE 186
Erica albertyniae E.G.H. Oliver sp. nov. (ms) 186
Erica ampullacea Curtis 188
Erica berzelioides Guth. & Bol. 188
Erica bodkinii Guth. & Bol. 188
Erica bruniades Guth. & Bol. 188
Erica bruniifolia Salisb. 188
Erica calcareophila E.G.H. Oliver 188
Erica casta Guth. & Bol. 190
Erica cerinthoides L. 190
Erica coccinea L. 190
Erica colorans Andr. 190
Erica corifolia L. 190
Erica discolor Andr. 190
Erica filipendula Benth. 192
Erica globulifera Dulfer 192
Erica grisbrookii Guth. & Bol. 192
Erica imbricata L. 192

Erica irbyana Andr. 192
Erica irregularis Benth. 194
erica-leaf spoon 112
Erica lineata Benth. 194
Erica longiaristata Benth. 194
Erica longifolia Ait. 194
Erica mariae Guth. & Bol. 194
Erica multumbellifera Berg. 196
Erica nudiflora L. 196
Erica oblongiflora Benth. 196
Erica occulta E.G.H. Oliver 196
Erica placentiflora Salisb. 196
Erica plukenetii L. 198
Erica propinqua Guth. & Bol. 198
Erica regia Bartl. 198
Erica rhopalantha Dulfer. 198
Erica scytophylla Guth. & Bol. 198
Erica sessiliflora L.f. 198
Erica shannonea Andr 200
Erica spectabilis Klotzsch ex Benth 200
Erica tenella Andr. 200
Erica vernicosa E.G.H.Oliver sp.nov. (ms) 200
Erica vestita Tunb. 200
Erica vogelpoelii H.A. Bak. 200
Eriocephalus paniculatus (Cass.) 238
Euchaetis 164
Euchaetis burchellii Dummer 164
Euchaetis longibracteata Schltr. 164
Euchaetis meridionalis I. Willams 164
Euclea racemosa Murray 204
Eucomis regia (L.) L'Herit. 64
EUPHORBIACEAE 168
Euphorbia erythrina Link 168
Euphorbia tuberosa L. 168
Euryops abrotanifolius (L.) DC. 238
Euryops linearis Harv. 238
ewwa–trewwa 98

FABACEAE 142
Falkia repens L. f. 208
false dodder 134

Felica aculeata Grau 240
Felica amoena (Sch. Bip.) Levyns subsp **latifolia** 240
Ferraria crispa Burm. 76
Ferraria latifolia Grau 240
Ficinia filiformis (Lam.) Schrad. 46
Ficinia praemorsa Nees 46
Ficinia truncata (Thunb.) Schrad. 46
fire lily 70
five–fingers 72
fluitjiesriet 42
fluted spiderhead 112
fluweeltjie 90, 220
fonteinbos 150
fonteinertjiebos 150
foxtail grass 50
freesia 76
Freesia elimensis L. Bol. 76
Freesia leichtinii Klatt 76
froetang 90
fyngras 38

galjoenblom 228
Gazania pectinata (Thunb.) Spreng 240
geldbeursie 60
geelbos 102, 108
geelgifbossie 252
geelkalkoentjie 86
geelkatstert 58
geelkoppie 143
geelmagriet 238, 250
geelsterrietjie 72
geelstompie 110
geelsuring 156
geel tjinkerintjee 64
geel viooltjie 54
Geissorhiza heterostyla L. Bol. 78
Geissorhiza inflexa (Delaroche) Ker–Gawl. 78
Geissorhiza ovata (Burm. f). Aschers. & Graebn. 78
geldbeursie 60

geneesbossie 176
GENTIANACEAE 206
GERANIACEAE 152
Geranium incanum Burm. f. var
   incanum 152
gifbol 68
Gladiolus 78
Gladiolus abbreviatus Andr. 78
Gladiolus brevifolius Jacq 78
Gladiolus bullatus Thunb. ex G.J. Lewis 80
Gladiolus carinatus Ait 80
Gladiolus carneus Delaroche 80
Gladiolus cunonius (L.) Gaertn. 80
Gladiolus floribundus subsp *miniatus* 84
Gladiolus debilis Ker Gawler 80
Gladiolus gracilis Jacq 82
Gladiolus guthriei F. Bol. 82
Gladiolus inflexus (in ed. Goldblatt & Manning 86
Gladiolus liliaceus Houtt. 82
Gladiolus maculatus Sweet 82
Gladiolus meridionalis Lewis 82
Gladiolus miniatus Eckl. 84
Gladiolus pillansii G.J. Lewis 84
Gladiolus punctulatus Schrank 84
Gladiolus rogersii Baker 84
Gladiolus stefaniae Oberm. 84
Gladiolus tenellus Jacq. 86
Gladiolus teretifolius Goldbl. & De Vos 86
Gladiolus tristis L 86
Gladiolus vaginatus Bolus f. 86
Gladiolus variegatus Goldblatt & Manning 86
glashout 100
Glottiphyllum depressum (Haw.) N.E. Br. 130
Gnidia albicans Meissner 180
Gnidia pinifolia L. **180**
Gnidia vesciculosa Eckl. & Zeyh. es Meisn. 180
Gnidia viridus Berg. 180
golden orchid 96

gonnabas 182
goue sterretjie 72
gousblom 230, 240
granny bonnet 96
green/bearded disa 96
green heath 198
green snake–stem pincushion 116
green wood orchid 98
groentaaiheide 196
groenviooltjie 62
groot bruin afrikaner 82
GRUBBIACEAE 124
Grubbia rosmarinifolia Berg. 124

HAEMODORACEAE 66
Haemanthus sanguineus Jacq. 70
Haemanthus coccineus 70
Hakea gibbosa (Sm.) Cav. 100
hangertjie 198
Harveya sp 216
Harveya capensis 216
Harveya purpurea 216
Hebenstretia cordata L. 222
Helichrysum chlorochrysum DC. 242
Helichrysum crispum (L.) D. Don 242
Helichrysum dasyanthum (Willd.) Sweet 242
Helichrysum paniculatum 244
Helichrysum retortum (L.) Willd. 242
Helichrysum teretifolium (L.) D. Don 242
Helichrysum vestitum 244
Heliophila macra Schltr. 134
Heliophila subulata Burch. ex DC. 134
Helipterum argyropsis 244
Helipterum canescens 244
Hellmuthea membranacea (Thunb.) R. Haines & Lye 46
Hermannia concinnifolia Verdoorn 176
Hermannia ternifolia Presl ex Harv. 176
Hermannia trifoliata L. 176
Herschelianthe lugens (H. Bol.) 96
Herschelianthe purpurascens (H. Bol.) 96

heuningblom 214
heuningbossie 234
Hibiscus trionum L. 176
Homeria bulbillifera G.J. Lewis 88
Homeria galpinii L. Bol. 88
Homoglossum abbreviatum 78
Homoglossum muirii 86
hondegesiggie 174
hondeoor 136
hongerblom 252
hottentots kooigoed 242
hottentotskool 58
hottentotsvy/fig 126
HYACINTHACEAE (Liliaceae) 60
Hyobanche sanguinea L. 216
Hyparrhenia hirta (L.) Stapf 40
Hypodiscus argenteus (Thunb.) Mast. 50
Hypodiscus willdenowia (Nees) Mast. 50
HYPOXIDACEAE 72

IRIDACEAE 74
Indigofera brachystachya (DC.) E. Mey. 148
inkplant 216
Ischyrolepis capensis (L.) Linder 52
Ischyrolepis eleocharis (Nees) Linder 52
Ischyrolepis leptoclados (Mast.) Linder 52
Ixia micranda Bak. 90

Jamesbrittenia albomarginata (Hilliard) 218
Jamesbrittenia calciphila (Hilliard) 218
Jamesbrittenia stellata (Hilliard) 218
Jordaniella dubia (Haw.) H.E.K. Hartm. 130
JUNCACEAE 58
Juncus kraussii Hochst. 58

kalkklipvygie 128
kalkoentjie 80
kalossies 90
kammetjie 76

kandelaar 68
kaneelbol 154
kaneelpypie 82
kanferblaar 152
kanferboom 240
kankerbos 150
kannetjies 208
kapokblom 66
kapokbossie 238, 240
katnaels 216
katpisbossie 214
katstert 58
kêr kêr 192, 196
kers(ie)hout 172
keurtjie 148
kiepersol 182
klamboegoe 162
kleinaalwyn 60
klein bruin afrikaner 82
kleinuintjie 88
klimop 208, 228
klipblom 148
klipdagga 214
klokkie 194
klokkiesheide 196
knikkertjie 90
Kniphofia uvaria (L.) Oken 60
knoppiesbos 104
Knowltonia anemonoides H. Rasm. 132
kolkol 138
kommetjieteewater 160
konings kandelaar 68
kooigoed 242
koraalbrak/bos 124
kouterbossie 230
kraaibessie 170
kraaltolbos 106
kransaalwyn 60
kroestaaibos 170
kruiphout 106
krulkransie 64
kweek 38

Lachenalia bulbifera (Cyr.) Engl. 62
Lachenalia contaminata Ait. 62
Lachenalia muirii W.F. Barker 62
Lachenalia rubida Jacq. 62
Lachnaea aurea Eckl. & Zeyh 182
Lachnospermum imbricatum (Berg.) Hilliard 250
lady's hand 72
lakpypie 94
LAMIACEAE 214
Lampranthus amabilis L. Bol. 128
Lanaria lanata (L.) Dur. & Schinz 66
langblaar 142
Lapeirousia pyramidalis (Lam.) Goldbl. 92
large painted lady 80
LAURACEAE 134
laventelbossie 250
lemoenbessie/doring 208
Leonotis leonuris (L.) R. Br. 214
lepelhout 172
Leucadendron 102
Leucadendron coniferum (L.) Meisn. 102
Leucadendron elimense Phill. subsp. elimense 104
Leucadendron laxum I. Williams 104
Leucadrendon linifolium (Jacq.) R.Br. 104
Leucadendron meridianum I. Williams 102
Leucadendron modestum I. Williams 102
Leucadendron muirii Phill. 106
Leucadendron platyspermum N. Br.106
Leucadendron salicifolium (Salisb.) I. Williams 108
Leucadendron salignum Berg. 106
Leucadendron stelligerum I. Williams 108
Leucodendron teretifolium (Andr.) I. Williams 108
Leucadendron xanthoconus (Kuntze) K. Schum. 110
Leucospermum 114

Leucospermum cordifolium (Salisb. ex Knight) Fourc. 114
Leucospermum cuneiforme (Burm. f.) Rourke 114
Leucospermum fulgens Rourke 114
Leucospermum heterophyllum
Leucospermum hypophyllocarpodendron subsp. hypophyllocarpodendron (l.) Druce 116
Leucospermum heterophyllum (Thunb.) Rourke 116
Leucospermum patersonii Phill. 114
Leucospermum pedunculatum Klotzsch 116
Leucospermum prostratum (Thunb.) Stapf 116
Leucospermum truncatulum (Salisb. ex Knight) Rourke 116
Leucospermum truncatum (Buek ex Meisn.) Rourke 114
lidjiesbos 124
Lightfootia rigida Adamson 226
limestone conebush 102
limestone pagoda 110
limestone pincushion 114
limestone sugarbush 120
Limonium anthericoides (Schltr.) R.A. Dyer 202
Limonium scabrum (Thunb.) Kuntze var scabrum 202
Limosella grandiflora Benth. 218
line–leaf conebush 104
Liparia splendens (Burm.f.) Bos. & De Wit 148
lippypie 80
LOBELIACEAE 228
Lobelia pubescens Dryand. ex Ait. 228
Lobelia setacea Thunb. 228
Lobelia tomentosa L. f. 228
Lobelia valida L. Bol 228
Lobostemon curvifolius Buek 210
Lobostemon lucidus {Lehm.} Buek 210

Lobostemon sanguineus Schltr 210
long–leaf sugarbush 120
long–stalk spiderhead 112
lotus lily 132
luise 114
luisiesbos 106
Lycium cinereum 212
Lyperia lychidea (L.) Druce 220

maagpynbossie 100, 152
Maartlelie 68
MALVACEAE 176
Manulea tomentosa (L.) L. 220
March lily 68
Massonia pustulata Jacq. 64
Mastersiella digitata (Thunb.) Gilg–Ben. 52
matras sewejaartjie 244
Maytenus procumbens (L. f.) Loes. 172
melkhout boom 202
melkhoutbos 202
MENISPERMACEAE 134
Merxmuellera cincta (Nees) Conert 40
Merxmuellera stricta (Schrad) Conert 40
MESEMBRYANTHEMACEAE 126
Metalasia 246
Metalasia brevifolia (Lam.) Levyns 246
Metalasia calcicola Karis 246
Metalasia muricata (L.) D. Don 246
Metalasia pungens D. Don 246
Metalasia serrata Karis 246
Micranthus junceus (Bak.) N.E. Br. 92
Microloma sagittatum (L.) R. Br., 208
milkbush 202
Mimetes 110
Mimetes cucullatus (L.) R, Br, 110
Mimetes saxatilis Phill. 110
minor sugarbush 118
mirting 202
misryer 70
moederkappie 96

Monadenia bracteata (Swartz) Dur. & Schinz 98
Monsonia emarginata (L. f.) L'Herit 152
Moraea 88
Moraea fugax (Delaroche) Jacq 88
Moraea neglecta G.J. Lewis 88
Moraea tripetala L.f. 88
mountain bluebell 80
Muraltia collina Levyns 166
Muraltia satureoides DC. var **satureoides** 166
Myrica cordifolia L. 100
Myrica quercifolia L. 100
MYRICACEAE 100
Myrisiphyllum asparagoides 64
MYRSINACEAE 202
Myrsine africana L. 202

naeltjies 180
Nebelia paleacea (Berg.) Sweet 138
needle-leaf conebush 108
Nemesia barbata (Thunb.) Benth. 220
Nemesia versicolor E. Mey. ex Benth. 220
nenta 148
nenta(bossie) 136
Nerine humilis (Jacq.) Herb. 68
nooienshaar 134
Nylandtia spinosa (L.) Dumort. 166
NYMPHACEAE 132
Nymphaea capensis 132
Nymphaea nouchali Burm. f. var **caerulea** (Sav.) Verdc 132

Oedera capensis (L.) Druce 240
Oedera genistifolia (L.) Anderb. & Bremer 240
ogies gousblom 240
Oktoberlelie 98
Olea capensis L. subsp **capensis** 204
OLEACEAE 204
Olea exasperata Jacq. 204
olifantsgras 40

265

Onixotis stricta (Burm.f.) Wijnands 58
ORCHIDACEAE 96
Ornithogalum dubium Houtt. 64
Ornithogalum thyrsoides Jacq. 64
Ornithoglossum viride (L.f.) Ait. 58
Orphium frutescens (L.) E. Mey. 206
Osyris compressa 122
Otholobium fruticans (L.) C.H. Stirton 150
Osteospermum fructicosum (L.) T. Norl. 248
Osteospermum subulatum DC. 248
Othonna dentata L. 248
Othonna filicaulis Jacq.248
Othonna quinquedentata Thunb.248
oumakappie 98
oval–leaf pincushion 116
OXALIDACEAE 156
Oxalis eckloniana Presl. var **sonderi** Salter 156
Oxalis luteola Jacq. 156
Oxalis pes-caprae L. 156
Oxalis polyphylla Jacq. 156
Oxalis purpurea L. 156

painted lady 80
palmiet 58
Paranomus 110
Paranomus abrotanifolius Salisb. ex Knight 110
Passerina ericoides L. 182
Passerina rigida Wikstr. 182
paasfeesblom 192
Pelargonium betulinum (L.) L'Herit. 152
Pelargonium capitatum (L.) L'Herit. 152
Pelargonium cucullatum (L.) L'Herit. 154
Pelargonium elegans (Andr.) Willd. 154
Pelargonium suburbanum Clifford ex Boucher subsp. **bipinnatifidum** (Harv.) Boucher 154
Pelargonium triste (L.) L'Herit 154
PENAEACEAE 178

Penaea mucronata L. 178
Pentaschistis eriostoma (Nees) Stapf 40
perdebiesie 90
perdebossie 250
perdekapok 66
perde–uintjie 88
Peucedanum galbanum (L.) Drude 184
Phaenocoma prólifera D. Don 250
Phragmites australis (Cav.) ex Steudel 42
Phylica 174
Phylica dodii N.E. Br. 174
Phylica ericoides L. 174
Phylica pubescens Ait. var **orientalis** Pillans 174
Phylica purpurea Sond. 174
Phylica selaginoides Sond. 174
Phylica stipularis L. 174
pink orchid 98
pineapple flower 64
pisgoed(bossie) 168
PLANTAGINACEAE 222
Plantago crassifolia Forssk. 222
plakkie 136
platdoring 184
plate–seed conebush 106
platy 106
Plexipus cernuus (L.) R. Fernandes 210
ploegtyd blommetjie 72
PLUMBAGINACEAE 202
POACEAE 38
poeierkwassie 230
Podalyria biflora Lam. 148
Podalyria cuneifolia Vent. 148
Polyarrhenia stricta Grau 250
POLYGALACEAE 166
Polygala myrtifolia L. 166
Polygala umbellata L. 166
poprosies 176
Port Jackson 142
Potberg pincushion 114
Potberg protea/sugarbush 118
Prenia vanrensburgii L. Bol. 130

Prionium serratum (L. f.) Drege ex E. Mey 58
Prismatocarpus brevilobus A. DC. 226
Prismatocarpus spinosus Adamson 226
PROTEACEAE 100
Protea 118
Protea aspera Phill. 118
Protea aurea (Burm. f.) Rourke subsp. **potbergensis** Rourke 118
Protea compacta R. Br. 120
Protea denticulata Rourke 118
Protea longifolia Andr.120
Protea obtusifolia Buek ex Meisn. 120
Protea pudens Rourke 118
Protea repens (L.) L. 120
Protea speciosa (L.) L. 118
Protea subulifolia (Salisb. ex Knight) Rourke 118
Protea susannae Phill. 120
Pseudopentameris macrantha (Schrad) Conert 42
Psoralea aphylla L. 150
Psoralea pinnata L 150
Pterocelastrus tricuspidatus (Lam.) Sond. 172
Pteronia incana (Burm.) DC. 250
Pterygodium catholicum (L.) Swartz 98
pyjamabos/bush 210
pynappelblom 64
pypsteelbos 140

Rafnia triflora Thunb. 150
rankluisie 116
rankmagriet 248
RANUNCULACEAE 132
red hot poker 60
reed bells 82
renosterbos(sie) 238
Restio multiflorus Sprengel 54
RESTIONACEAE 48
Restio triticeus Rottb. 54
Retzia capensis Thunb. 214

RETZIACEAE 214
RHAMNACEAE 174
Rhigiophyllum squarrosum Hochst. 224
Rhus 170
Rhus crenata Thunb. 170
Rhus glauca Thunb. 170
Rhus lucida L. 170
Rhus rosmarinifolia Vahl 170
Rhyticarpus difformis 184
rock hakea 100
Roella arenaria Schltr. 224
Roella incurva A. DC. 224
Roella rhodantha Adamson 224
roemenaggie 180
Romulea flava (Lam.) De Vos var **flava** 90
Romulea rosea L. Eckl.var **reflexa** (Eckl.) Beg 90
rooibeentjies 133
rooiblombossie 246
rooidagga 214
rooigras 44
rooihartjie 190
rooi inkblom 216
rooikanol 66
rooikrans 142
rooipypie 78, 94
rooisewejaartjie 244, 250
rooitolbos 102
rooistompie 110
rooi–trewwa 98
rooi viooltjie 62
rooiwortel 66,154
ROSACEAE 140
rose scented pelargonium 152
rosyntjiebos 170
rough–leaf conebush 102
rough–leaf sugarbush 118
Ruschia geminiflora (Haw.) Schwant. 128
RUTACEAE 160

Saltera sarcocolla (L.) Bullock 178
Salvia africana–lutea L. 214

sandkalossie 62
sandlelie 70
sandpypie 82
sandvygie 128
SANTALACEAE 122
SAPOTACEAE 202
Sarcocornia littorea (Moss) A.J. Scott 124
Satyrium carneum (Dryand.) Sims 98
Satyrium coriifolium Swartz 98
Scabiosa columbaria L. 222
SCROPHULARIACEAE 216
seaguarri 204
sea lavender 202
sea wheat 44
Sebaea aurea (L. f.) Roem. & Schult. 206
seepampoen 130, 230
seeroogblom 68
sekelbos 100
SELAGINACEAE 222
Selago aspera Choisy 222
Selago serrata Berg. 222
Senecio arenarius Thunb. 252
Senecio burchellii DC. 252
Senecio elegans L. 252
Senecio sophioides DC. 252
Serruria 112
Serruria bolusii Phill. & Hutch 112
Serruria elongata (Berg.) R. Br. 112
Serruria fasciflora Salisb. ex Knight 112
Serruria nervosa Meisn. 112
sewejaartjie 236, 244
sickle–leaf conebush 110
Sideroxylon inerme L. 202
sieketroos 184
Silene undulata Ait. 132
silky spoon 110
silver–ball conebush 106
silver–edge pincushion 114
Simocheilus purpureus (Berg.) Druce 186
sissie heath 188
skaamgesiggie 118
skaapbostee 150

skilpadbessie 166, 182
skilpadbos 158
skilpadbossie 102, 124
slakblom 134
slangbessiebos 212
slangbos 236
slangbossie 116
slanghout 204
slaptaaibos 170
slymbos 158
snotbel 198
snotblom 134
snowball 118
SOLANACEAE 212
Solanum quadrangularis Thunb. ex L.f. 212
soldier–in–the–box 60
sour fig 126
soutbossie 124
Sparaxis bulbifera (L.) Ker–Gawl. 90
Spatalla 110
Spatalla curvifolia Salisb. ex Knight 112
Spatalla ericoides Phill. 112
Spatalla squamata Meisn. 110
spekbossie 158
spekbroodbossie 158
speldekussing 114
spider orchid 96
Spiloxene aquatica (L.f.) Fourc. 72
Spiloxene capensis (L.) Garside 72
Spiloxene flaccida (Nel) Garside 72
spinnekopbossie 112
spinnekopblom 66, 76
spinnekoporgidee 96
Staavia radiata (L.) Dahl 138
Staberoha multispicula Pillans 54
Stachys aethiopica L. 214
statice 202
steekbos 246
Stenotaphrum secondatum (Walter) Kuntze 42
STERCULIACEAE 176

sterretjie 72
sterretjies 108
STILBACEAE 212
Stilbe ericoides (L.) L. 212
stink–blaar(leaf)protea/sugarbush 120
Stoebe plumosa (L.) Thunb. 236
stompie 110
strandbietou 232
strandblommetjie 252
strandsalie 214
strooiblommetjie 244
Struthiola argentea Lehm. 180
suikerbos 120
sundew 134
Sutera hispida (Thunb.) Druce 216
Sutera lychnidia 220
Sutera revoluta (Thunb.) Kuntze 216
suurkanol(pypie) 76, 94
suurvy 126
Sutherlandia frutescens (L.) R. Br. 150
swaartbaard 120
swartbal 106
swartwattel 142
Syncarpha argyropsis (DC.) B. Nord. 244
Syncarpha paniculata (L.) B. Nord. 244
Syncarpha canescens (L.) B. Nord. 244
Syncarpha vestita (L.) B. Nord. 244
Syndesmanthus articulatus (L.) Klotzsch 186

tandebossie 176
tandjies 118
Tarchonanthus camphoratus L. 240
TECOPHILACEAE 72
Tetragonia decumbens Mill. 124
Tetragonia herbacea L. 124
Tetraria bromoides (Lam.) Pfeiffer 44
Tetraria thermalis (L.) C.B. Cl 44
Thamnochortus insignis Mast. 56
Thamnochortus fraternus Pillans 56
Thamnochortus pellucidus Pillans 56
thatching grass 56

Themeda triandra Forssk 44
"The" pincushion 114
Thesium capitatum L. 122
Thesium fragile (Thunb.) Sonder 122
Thesium penicillatum A.W. Hill 122
Thinopyrum distichum (Thunb.) Loeve 44
Thoracosperma puberulum (Klotzsch) N.E. Br. 186
THYMELAEACEAE 180
tjêr-tjêr 196
tjinkerintjee 64
tongblaar 218
tongblaarvygie 130
tonteldoek(blom) 230
tooth–leaf sugarbush 118
Trachyandra divaricata (Jacq.) Kunth 58
tregterheide(heath) 190
trident pincushion 116
trilheide/heath 200
Tritonia deusta (Ait,) Ker Gawler 90
Tritoniopsis antholyza (Poir.) Goldbl. 92
Tritoniopsis apiculata (F.Bol.) G.J. Lewis 92
Tritoniopsis dodii (G.J. Lewis) G.J. Lewis 92
tulp 88

uintjiestulp 88
Ursinia paleacea (L.) Moench 250

vaalstompie 138
vaaltol 138
varkblom 46
veerkoppie 174
veldskoenblaar 70
VERBENACEAE 210
vingersuring 156
VIOLACEAE 178
Viola decumbens L. f. 178
VISCACEAE 122
Viscum capense L. f. 122
vlakteheide/heath 190

vlei-aandblom 86
vleiblom 58
vleiblommetjie 92
vlei daisy 236
vleipypie 86
vleirosie 104
vlei–uintjie 72
vlieëbos(sie) 178, 202
voëlent 122
volstruisvygie 126
vrouebossie 152
vuurpyl 60

waboom 114
Wachendorfia paniculata Burm. 66
Wachendorfia thyrsiflora Burm. 66
Wahlenbergia calcarea (Adamson) Lammers 226
Wahlenbergia capensis (L.) A. DC. 226
wart–stemmed pincushion 114
wasbessie 100
waspypie 94
waterblommetjie 38
water lily 132
watersterretjie 72
wateruintjie 38
Watsonia 94
Watsonia aletroides (Burm. f.) Ker–Gawl. 94
Watsonia coccinea Herb. ex Bak. 94
Watsonia fergusoniae L. Bol. 94
Watsonia laccata (Jacq.) Ker–Gawl. 94
Watsonia meriana (L.) Mill. 94
waxberry 100
waxcreeper 208
weeskindertjies 220
white-stalked spoon 112
white trailing pincushion 116

wide mouth heath 200
wild anemone 132
wild cineraria 252
wild hyacinth 62
wild pink 132
wild rosemary 238
wild scabious 222
wild violet 178
wilde–angelier 132
wildegaansie 150
wilde dagga 214
wilde/wild lobelia 228
wildemalva 154
wilde mirting 202
wilde–olyfboom 204
wildetabak 132
witbergpypie 70
witmagriet 234, 236
wit sewejaartjie 236, 244
wit viooltjie 64
wolkop 120
wolwekos 216

Xeroplana zeyheri Briq. 212

yellow chinkerinchee 64
yellow trailing pincushion 116
ysterhout 200

Zantedeschia aethiopica (L.) Spreng 46
Zaluzianskya villosa (Thunb.) F.W. Schmidt 220
ZYGOPHYLLACEAE 158

Zygophyllum flexuosum Eckl & Zeyh. 158
Zygophyllum fulvum L. 158
Zygophyllum morgsana L. 158
Zygophyllum sp nov 158

About the Botanical Society of South Africa

Founded in 1913 at the same time as Kirstenbosch Botanic Gardens, the Botanical Society aims to interest the people of South Africa and other countries in the National Botanic Gardens. We also aim to educate members of the public in the cultivation, conservation and awareness of our unique indigenous flora.

ARE YOU A MEMBER?
The Botanical Society of South Africa is one of the largest, most effective organisations working to safeguard our veld and flora. If you are not already a member we invite you to join. There is something for everyone in the Society's wide range of activities, from hikes and walks to illustrated lectures, tours and conservation activism. Members receive the colourful and informative "Veld & Flora" magazine, free seeds of your choice annually from the Kirstenbosch seed list, as well as free admission to all the national botanic gardens in South Africa.

By joining the Society you support those members who are willing to invest their time and expertise to protect our natural heritage for this and future generations. We need your membership and support. To join, please contact the Executive Secretary, Botanical Society of South Africa, Kirstenbosch, Claremont 7735 RSA or telephone Cape Town (021) 797-2090.

If you have enjoyed the wild flower guides in this series and wish to support us in our programme of producing guides for other areas of South Africa, donations may be sent to the Botanical Society of South Africa for its publication programme. Any donations or bequests made to the Botanical Society or its Flora Conservation Committee are free of donations and estate duty tax.

Plants are protected by Ordinance 19 of 1974 as amended which prohibits the picking of any plant within 90 m of the middle of the road, or the picking of any plant without the written permission of the landowner, or the picking of any species that are proclaimed endangered or protected without the necessary permits. Permits are obtainable on written application from:

The Chief Director
Cape Nature Conservation
Private Bag X9086
8000